# ALGEBRA

*Math Tutor Lesson Plan Series*

# Book 1

iGlobal Educational Services

# iGlobal

## Educational Services

Believe. Inspire. Transform.

To order, contact iGlobal Educational Services, PO Box 94224, Phoenix, AZ 85070

Website: www.iglobaleducation.com

Fax: 512-233-5389

ISBN-13: 978-1-944346-64-5

Printed in the United States of America.

# Algebra Book 1

## Contents

# Introduction

Tutoring is beginning to get the respect and recognition it deserves. More and more learners require individualized or small group instruction whether it is in the classroom setting or in a private tutoring setting either face-to-face or online.

This lesson plan book is part of the series "Math Tutor Lesson Plan" Series. It is conceived and created for tutors and educators who desire to provide effective tutoring either in person or online in any educational setting, including the classroom.

## Inside This Lesson Plan Book

This *Algebra: Math Tutor Lesson Plan Series* book provides appropriate practice during tutoring sessions for learners for both face-to-face and online tutoring sessions focused on topics in Pre-Calculus.

The goal of the *Algebra: Math Tutor Lesson Plan Series* book is to support all types of tutors. Also, this book is to support teachers who want to provide in-class tutoring to their students in either an individualized or small group tutoring setting. Lastly, this book is also for teachers who are providing math intervention either individually or in small group tutoring sessions either face to face or online so that they can select the specific lesson plan to address the learner's math learning needs.

## How to Use This Lesson Plan Book

iGlobal Educational Services, in collaboration with, Dr. Alicia Holland-Johnson, Tutor Expert and Consultant, created this tutoring resource to help with designing effective tutoring instruction for tutors and teachers who desire to provide in-class tutoring sessions.

These specific lessons were selected based upon field-tested experiences with learners who had learning needs over the years in these specific areas in mathematics. We have provided learning objectives and specific topics covered in each tutoring session so that you can align them with your state's specific standards or adapted standards. For overseas tutors, you can follow suite and align the lesson objectives to specific educational standards required in your country.

These lesson plans should be used to supplement a strong and viable curriculum that encourages differentiation for all diverse learners. They can be used in individual or small group tutoring sessions conducted face-to-face or online in any educational setting, including the classroom.

## Organization of the Lesson Plan Book

Rather than provide a specific "curriculum" to follow, *Algebra: Math Tutor Lesson Plan Series* book provides a blueprint to design effective tutoring lessons that are aligned with the "*Dr. Holland-Johnson's Session Review Framework*". Tutor evaluators and coaches are able to analyze tutoring sessions and coach tutors when utilizing the "*Dr. Holland-Johnson's Lesson Plan Blueprint for Tutors*". In each lesson plan, learners have an opportunity to focus on real-world connections, vocabulary, and practice the math concepts learned in the tutoring sessions in the appropriate amounts to learn and retain the content knowledge. Tutors will have an opportunity to provide direct and guided instruction, while learners practice concepts on their own during independent instruction.

Each lesson plan comes with a mini-assessment pertaining to the math concepts learned in the specific tutoring session. Depending on the learner's academic needs, the tutor or teacher will deem when it is appropriate to administer the mini-assessment. For online tutoring sessions or as an online option to take the mini-assessment, tutors and teachers can upload these mini-assessments to be completed online in their choice of an online assessment tool.

# Lesson 1
# Linear and Absolute Value Equations

## Lesson Description

This lesson is designed to help students solve linear and absolute value equations. Please be sure to utilize the questions to help spark student engagement and cover the vocabulary that is associated with this specific tutoring session. For your own knowledge, sample responses have been provided to guide you as well.

## Learning Objective(s)

In today's lesson, the learners will solve linear and absolute value equations in 3 out of 4 trials with at least 75% or higher accuracy.

# Introduction

When was the last time that you checked your body temperature? It is common knowledge that a person's normal body temperature is supposed to be 98.6 degrees. If the temperature is any higher than that, then physicians deem that an individual may have a fever and possibly sick.

When was the last time you traveled? How far was it, in distance? These are questions that can be answered by using absolute value equations to represent the relationships that are presented to solve for the minimum and maximum numbers for the solutions.

## Questions to Engage Students

➤ Do you people know what is a linear equation and how is it identified as linear?

➤ Can you give some real life example of a linear equation?

➤ What is meant by solving an equation?

➤ What does absolute value means?

➤ Write an example of absolute value equation?

## Connect Learning Objective(s) Student's Lives

**A)** Linear Equations are used in:

➤ Temperature Conversion $C = 5/9\ (F-32)$

➤ Exchange Rates

➤ Finding Distance when rate and time is given or vice versa.

➤ Cell Phone Plans

➤ Calculating Net Pay, Profit, Loss etc.

**B)** Absolute value equations are used:

➤ To calculate the range of values when some tolerance in manufacturing or else is given.

# Specific Vocabulary Covered

### Linear Equation

A linear equation is a degree '1' equation. It means the highest exponent of the variable of equation is '1'. A common linear equation in two variables can be written as,

$$y = mx + b$$

### Absolute Value

Absolute value is represented as $|a|$. It always evaluates to a positive number.

$$|a| = |-a| = a$$

### Absolute Value Equation

An equation that contain an absolute value is called an absolute value equation. For example,

$$|ax + b| = c \text{ where } c > 0$$

# Direct & Guided Instruction: Modeling For You and Working With You

## Solving Linear Equations

➤ The idea of solving linear equations is to isolate the variable at one side of the equation.

➤ To isolate the variable, use inverse operation that is done to the variable, on both sides of the equation to keep the equation balanced.

➤ For example if 4 is added to the variable $x$, subtract 4 on both sides of the equation.

## Number of Possible Solutions for Linear Equations

When a linear equation is solved, three cases can be yielded.

| Case I | Case II | Case III |
|---|---|---|
| $x = a$ (constant) | $a = b$ | $a = a$ |
| Linear equation has one Solution. | Linear equation has no Solution. | Linear equation has infinite Solutions |

Where $a$ and $b$ are constants.

☞ **Solve the linear equation.**

$$2x - 13 = 5 - 4x$$

### Teacher's Questions

➤ What will be the very first step we perform?

➤ In which variable this linear equation is?

➤ What is meant by isolating the variable?

➤ How to isolate the variable '$x$'?

➤ How to verify if the value of '$x$' is correct?

**Solution**

Solve the linear equation.

$2x - 13 = 5 - 4x$

Add 13 on both sides

$2x - 13 + 13 = 5 - 4x + 13$

$2x = 18 - 4x$

To isolate the variable on one side, add $4x$ on both sides,

$2x + 4x = 18 - 4x + 4x$

$6x = 18$

Divide by 6 on both sides

$6x/6 = 18/6$

What is the value of $x$?

$x = 3$

## Steps to Solve Absolute Value Equations

➤ Isolate the absolute value to one side of the equation.
➤ Check if the number on the other side is positive or negative.
  ➤ If negative, there is no solution to the equation.
  ➤ If positive, then proceed further.
➤ Split the absolute value equation to two cases, one for positive value of the absolute value and one for negative. For example,

If equation is |x+1| = 2, split it into,

$(x + 1) = 2$ and

$-(x + 1) = 2$

➤ Solve both the cases.
➤ Check all the roots by putting in the original equation to filter out any extraneous root.

## Number of Possible Solutions for Absolute Value Equations

When Absolute equations are solved, three cases are yielded.

| Case I | Case II | Case III |
|---|---|---|
| If $|ax + b| = 0$ | If $|ax + b| = -c$ | If $|ax + b| = c$ |
| Equation has one solution. | Equation has no Solution. | Equation has two solutions. |

☞ **Solve the absolute value equation.**

$$|5x - 1| = 2x + 3$$

## Teacher's Questions

➤ What should be the first step while solving an absolute value equation?

➤ How to split the absolute value equation in to two cases, positive and negative?

➤ What if the root of an equation does not satisfy the original equation?

## Solution

Solve the absolute value equation.

$$|5x - 1| = 2x + 3$$

Split absolute value equation into two cases:

| Case 1 | Case II |
|---|---|
| $5x - 1 = 2x + 3$ | $-(5x - 1) = 2x + 3$ |
| Add 1 on both sides: | $-5x + 1 = 2x + 3$ |
| $5x - 1 + 1 = 2x + 3 + 1$ | Subtract 1 from both sides: |
| $5x = 2x + 4$ | $-5x + 1 - 1 = 2x + 3 - 1$ |
| Subtract $2x$ from both sides: | $-5x = 2x + 2$ |
| $5x - 2x = 2x + 4 - 2x$ | Subtract $2x$ from both sides: |
| $3x = 4$ | $5x - 2x = 2x + 2 - 2x$ |
| Divide by 3 on both sides: | $7x = 2$ |
| $x = 4/3$ | Divide by $-7$ on both sides: |
| To verify the solution, put $x = 4/3$ in | $x = -2/7$ |
| $|5x - 1| = 2x + 3$ | To verify the solution, put $x = -2/7$ in |
| $|5(4/3) - 1| = 2 (4/3) + 3$ | $|5x - 1| = 2x + 3$ |
| $17/3 = 17/3$ | $|5(-2/7) - 1| = 2 (-2/7) + 3$ |
| | $17/7 = 17/7$ |

Therefore, $x = -2/7, 4/3$

☞ **The sum of two consecutive odd numbers is 28. Find the numbers?**

## Teacher Questions

➤ How to assume an odd number in the form of a variable '$x$'?

➤ What will be the other odd number in form of '$x$'?

➤ How to write a linear equation for this question?

➤ At the end you got the value for '$x$', how will you evaluate the consecutive odd numbers?

➤ Is there any way to verify that the odd numbers we got at the end are correct?

➤ The sum of two consecutive odd numbers is 28. Find the numbers?

## Solution

Let first odd number = $2x + 1$

Second odd number = $2x + 3$

According to the question statement,

$$(2x + 1) + (2x + 3) = 28$$
$$4x + 4 = 28$$

Subtract 4 on both sides:

$$4x + 4 - 4 = 28 - 4$$
$$4x = 24$$

Divide by 4 on both sides:

$$4x/4 = 24/4$$
$$x = 6$$

Therefore, first odd number = $2(6) + 1 = 13$

Second odd number = $2(6) + 3 = 15$

☞ **The temperature averaged 43 degrees last month. Temperatures were about 8 degrees warmer or colder. Write and solve an absolute value equation to represent the situation?**

## Teacher Questions

➤ What is the difference of temperatures given?

➤ How will you write an absolute value equation?

➤ Split the equation into two cases for positive and negative value of the absolute value?

➤ The temperature averaged 43 degrees last month. Temperatures were about 8 degrees warmer or colder. Write and solve an absolute value equation to represent the situation?

## Solution

$|x - 43| = 8$

| Case I | Case II |
|---|---|
| $x - 43 = 8$ | $-(x - 43) = 8$ |
| Add 43 on both sides: | $-x + 43 = 8$ |
| $x - 43 + 43 = 8 + 43$ | Subtract 43 on both sides: |
| $x = 51$ | $-x + 43 - 43 = 8 - 43$ |
|  | $-x = -35$ |
|  | $x = 35$ |

# Video Suggestions

Please conduct a search on either YouTube or Teacher Tube to find appropriate videos for this lesson. Below are some suggested title searches:

➤ Possible Solutions for Linear Equation

➤ Possible Solutions for Absolute Value Equation

➤ Solving Linear Equation

➤ How To Solve Absolute Value Equation

# Independent Instruction: Working on Your Own

## Questions

☞ **Solve the absolute value equation.**

$$|2x - 5| + 1 = 18$$

☞ **Jane is four times as old as Steve. After 12 year Jane will be twice as that of Steve. How old Jane and Steve are now?**

☞ **During the packaging process, a machine fills 400g of tea in packets. The machine always fills 5g less or more than the desired weight. Find the minimum and maximum weight of the packets?**

| Solution |
| --- |

**1.** $|2x - 5| + 1 - 1 = 18 - 1$

$|2x - 5| = 17$

Split the absolute value equation in to two cases:

| Case I | Case II |
| --- | --- |
| $2x - 5 = 17$ | $-(2x - 5) = 17$ |
| Add 5 on both sides: | $-2x + 5 = 17$ |
| $2x - 5 + 5 = 17 + 5$ | Subtract 5 from both sides: |
| $2x = 22$ | $-2x + 5 - 5 = 17 - 5$ |
| Divide by 2 on both sides: | $-2x = 12$ |
| $2x/2 = 22/2$ | Divide by $-2$ on both sides: |
| $x = 11$ | $-2x/-2 = 12/-2$ |
| | $x = -6$ |

**2.** Steve's age $= x$     Jane's age $= 4x$

After 12 years

Steve's age $= x + 12$     Jane's age $= 4x + 12$

According to the question statement,

$4x + 12 = 2(x + 12)$

$4x + 12 = 2x + 24$

Subtract 12 from both sides:

$4x + 12 - 12 = 2x + 24 - 12$

$4x = 2x + 12$

Subtract $2x$ from both sides:

$4x - 2x = 2x + 12 - 2x$

$2x = 12$

Divide by 2 on both sides:

$2x/2 = 12/2$

$x = 6$

Steve's age $= x = 6$ years

Jane's age $= 4x = 4(6) = 24$ years

**3.** $|x - 400| = 5$

Split the absolute value equation into two cases:

| Case I | Case II |
|---|---|
| $x - 400 = 5$ | $-(x - 400) = 5$ |
| Add 400 on both sides | $-x + 400 = 5$ |
| $x - 400 + 400 = 5 + 400$ | Subtract 400 from both sides: |
| $x = 405$ | $x + 400 - 400 = 5 - 400$ |
| | $-x = -395$ |
| | $x = 395$ |

Therefore, minimum and maximum weight of packets of tea is 395g and 405g.

# Mini-Assessment

**1.** The roots of the equation $|2x-1|-3 = 12$ are:

    **A.** 8, 7       **B.** −8, 7       **C.** 8, −7       **D.** −8, −7       **E.** 8, −8

**2.** The equation $5x + 19 = -21 - 5x$ evaluates to:

    **A.** $x = 0$       **B.** $x = -4$       **C.** $x = 4$       **D.** $x = -1/4$       **E.** $x = 1/4$

**3.** The equation $|5x| + 17 = 5$ evaluates to:

    **A.** $x = 12/5, 12/5$

    **B.** $x = 12/5, -5/12$

    **C.** $x = 12/5, -12/5$

    **D.** $x = 5/12, -5/12$

    **E.** No Solution

**4.** For the following linear equation, state how many number of solutions will be there?

$$5x - 8 = 3x + 5 + 2x$$

**5.** John makes \$35 on each TV set he sells. If his daily expenses are \$250, how many TV sets he needs to sell to make a minimum profit?

**6.** A thermometer has a tolerance of 1.5 degrees. If on some specific day the temperature is 23 degrees, what are the minimum and maximum values of temperature in degrees, the thermometer will show?

**7.** Evaluate the following Absolute value equation.

$$|3x + 15| - 9 = 0$$

# Mini-Assessment Answers and Explanations

1. C

2. B

3. E

4. No Solution

5. 7 TV Sheets

6. 21.5, 24.5

7. $x = -2, -8$

## Lesson Reflection

- ➤ Writing linear equations and Absolute value equations with the help of given question statement.

- ➤ Solving Linear Equation.

- ➤ Finding number of possible solutions of linear equation.

- ➤ Solving Absolute Value Equations.

- ➤ Finding number of possible solutions of absolute value equation.

- ➤ Filtering out extraneous solutions.

# Lesson 2
# Solving and Graphing Inequalities

## Lesson Description:

This lesson is designed to help students graph the solutions to a linear inequality in two variables as a half-plane (excluding the boundary in the case of a strict inequality), and graph the solution set to a system of linear inequalities in two variables as the intersection of the corresponding half-planes. Please be sure to utilize the questions to help spark student engagement and cover the vocabulary that is associated with this specific tutoring session. For your own knowledge, sample responses have been provided to guide you as well.

## Learning Objective(s):

In today's session, the learners will solve and graph linear inequalities in 3 out of 4 trials with at least 75% or above accuracy.

## Introduction

Look around you! Inequalities are used all the time around us. For instance, when you are in the car as either a driver or passenger, what do you see? There are boatloads of speed limit signs posted on the highway. Have you thought about the minimum payments on a credit card bill? What about the number of text messages in which you send each month from your cell phone? All of these can be represented as mathematical inequalities. Therefore, it is extremely important to breakdown everyday situations so that you can interpret the language and apply inequalities to solve the problem and present a solution.

## Questions to Engage Students

➤ In your own word, how would to describe a linear equation.

➤ How would you differentiate between a linear equation and a linear inequality.

➤ How a linear equation or a linear inequality can be written in a slope intercept form.

$$y = mx + b$$

## Connect Learning Objective(s) Student's Lives

**A)** When we go into the store and we are looking at any item for that matter we think about the amount of money we have (we usually estimate our total) and then we calculate in our heads the largest amount of the items that we can buy with the money we have. It means that the total amount we spend should be less than or equal to the money we have in our pocket.

**B)** Inequalities are also used when we are participating in competitions, like we should attain greater than or equal to some certain percentage.

# Specific Vocabulary Covered

## Linear Inequality

Linear Inequality is an inequality of the form

$$y \leq mx + b, y \geq mx + b,$$

$$y > mx + b, y < mx + b$$

Where '$m$' and '$b$' are constants and '$x$' and '$y$' are variables.

## Test Point

$(0, 0)$ is selected as a test point to check, which side of the plotted line is the solution set of the inequality.

## Solid Line

If inequality is of the form $y \leq mx + b$ or $y \geq mx + b$, solid line is plotted for the corresponding linear equation $y = mx + b$.

## Dotted Line

If inequality is of the form $y < mx + b$ or $y > mx + b$, dotted line is plotted for the corresponding linear equation $y = mx + b$.

# Direct & Guided Instruction: Modeling For You and Working With You

## Steps to Graph an Inequality

➤ Change the linear inequality into linear equation.

➤ Graph a solid line, if the inequality has the ≤ or ≥ sign and graph a dotted if inequality has < or > sign.

➤ Pick a test point to find out which part needs to be shaded. (0, 0) is an easy ordered pair to test.

➤ If (0, 0 ) satisfies the inequality, shade the half plane where (0, 0) lies, otherwise shade the other half plane.

☞ **Solve and graph the inequality.**

$$y > 2x + 1$$

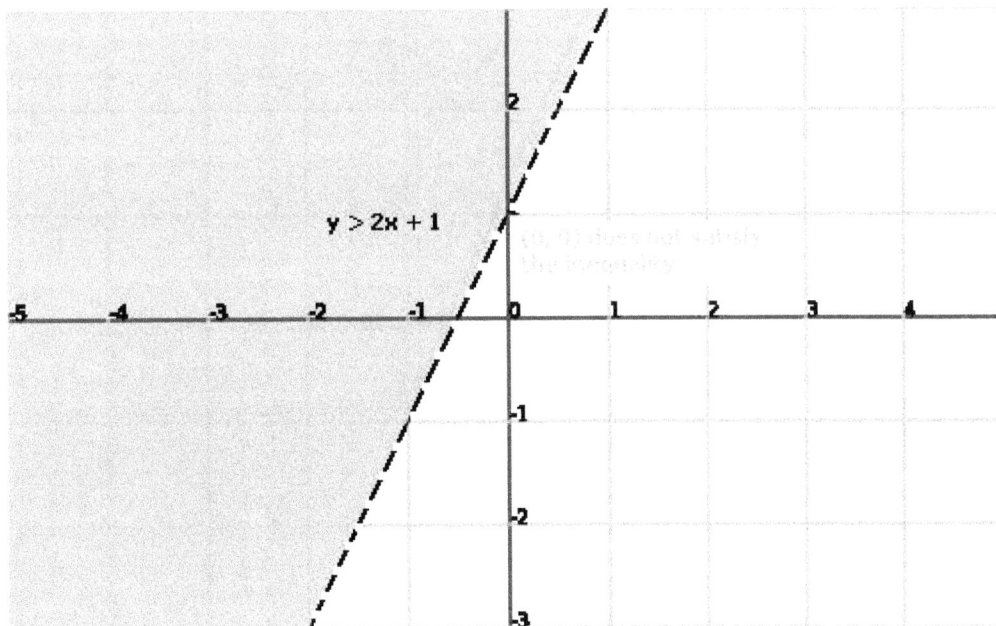

## Teacher's Questions

➤ How an inequality can be written in a slope intercept form.

➤ What is the idea of a test point.

➤ How to determine, which side of the line is to be shaded.

➤ What s the difference between a dotted line n a solid line.

➤ A man sells jackets 15$ each and shirts 10$ each. How many jackets and shirts must be sold to make a profit greater than or equal to 150$?

## Solution

If $x$ is number of jackets sold and $y$ is the number of shirts sold then

$15x + 10y \geq 150$

$3x + 2y \geq 30$

$2y \geq 30 - 3x$

$y \geq 15 - 1.5\,x$

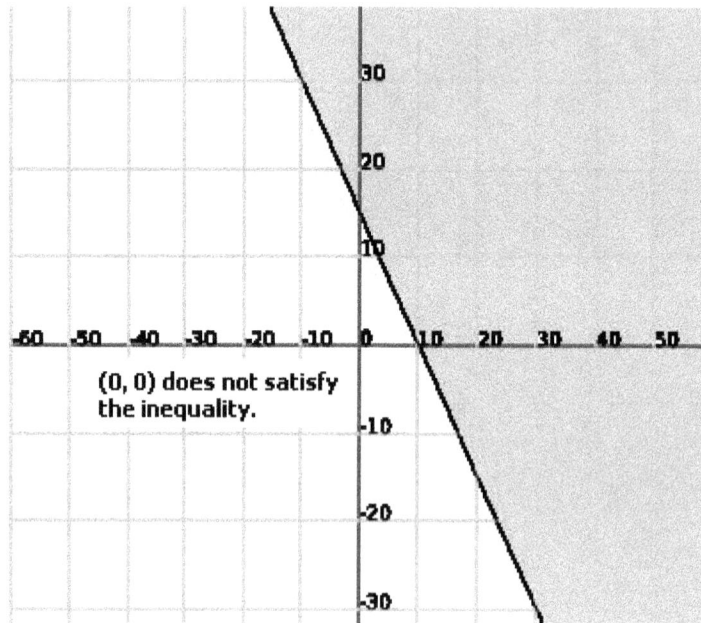

(0, 0) does not satisfy the inequality.

☞ **Solve and Graph** $x - y \leq 2$.

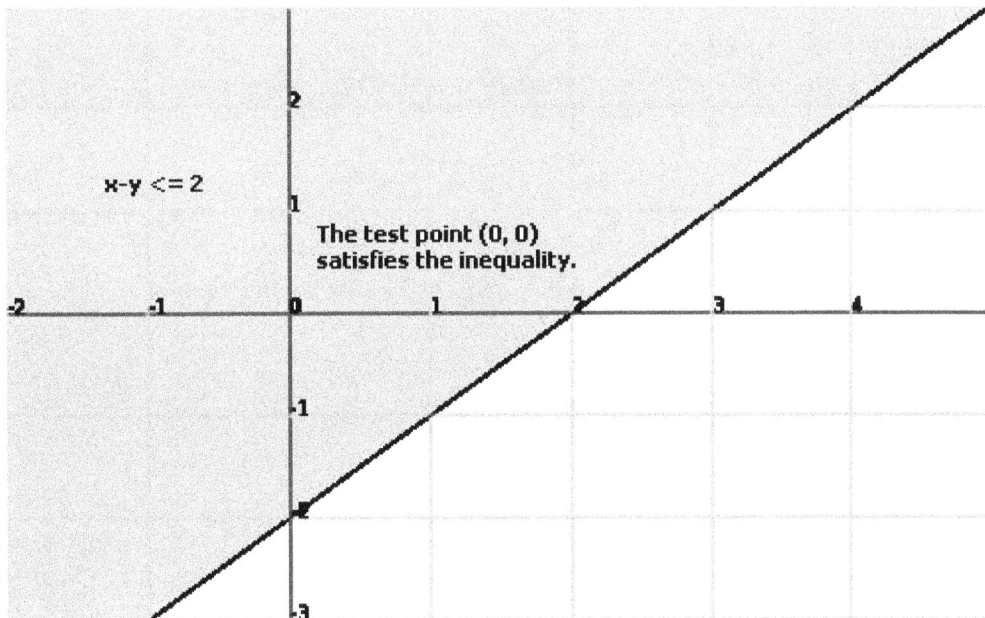

x-y <= 2

The test point (0, 0) satisfies the inequality.

## Teacher's Questions

➤ What will be the corresponding linear equation.

➤ How would you graph the corresponding linear equation.

➤ What type of line would it be? Dotted or Solid.

➤ How the test point is to be verified?

➤ If bananas cost 8$ per dozen and oranges cost 10$ per dozen. A person has 30$ in total, if he needs to buy 2 dozen bananas, what maximum number of oranges can he buy?

➤ Identify the variables.

➤ How the cost of '$x$' dozen bananas can be specified.

➤ How the cost of '$y$' dozen oranges can be specified.

➤ How to write the statement in the form of an inequality.

➤ If bananas cost 8$ per dozen and oranges cost 10$ per dozen. A person has 30$ in total, if he needs to buy 2 dozen bananas, what maximum number of oranges can he buy?

## Solution

Let number of bananas = $x$ dozen

Number of oranges = $y$ dozen

$$8x + 10y \leq 30$$

$$4x + 5y \leq 15$$

$$5y \leq 15 - 4x$$

$$y \leq -4/5\, x + 3$$

To find maximum number of oranges to buy, put $x = 2$ in the inequality,

$$y \leq -4/5\, x + 3$$

$$y \leq -4/5\, (2) + 3$$

$$y \leq -8/5 + 3$$

$$y \leq 7/5$$

$$y \leq 1.4$$

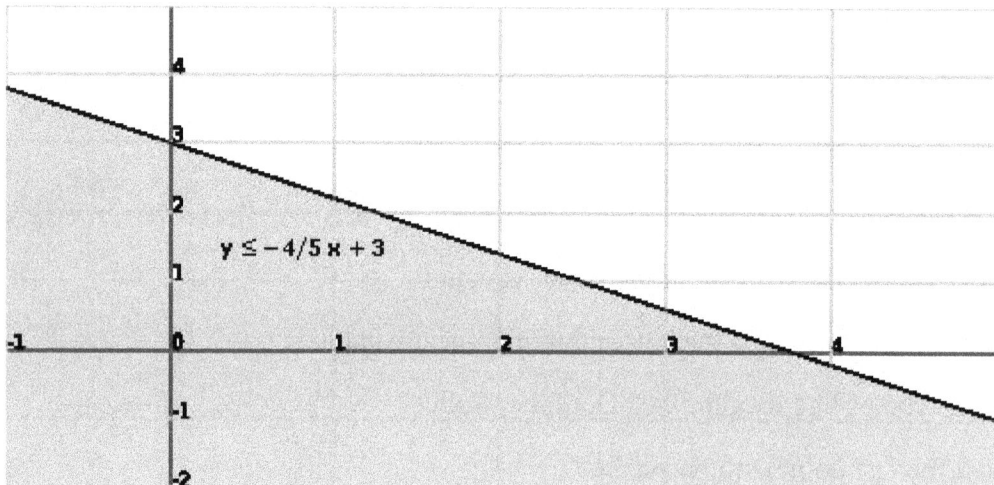

# Video Suggestions

Please conduct a search on either YouTube or Teacher Tube to find appropriate videos for this lesson. Below are some suggested title searches:

➤ Solving Linear Inequality

➤ How To Graph Linear Inequality

➤ Identifying Variables

➤ How To Graph Linear Equation

# Independent Instruction: Working on Your Own

## Questions

☞ Graph the inequality.

$$3y - 2x > 5$$

☞ If a car dealer has $1,408,000 available to purchase compact cars and sport utility vehicles. The compact car costs $11,000 and the sport utility vehicle costs $22,000. Write an inequality that models the different numbers of compact cars and sport utility vehicles that he could purchase.

☞ It costs 8$ to buy a Math Book and 10$ to buy a Science Book. If you have 90$ to spend, write an inequality to represent all possible combinations of Math and Science books. If you bought 7 Math books, what are the maximum number of Science books you can buy?

**1.** $3y > 2x + 5$

$y > (2/3) x + 5/3$

The line will be dotted because of the '>' sign.

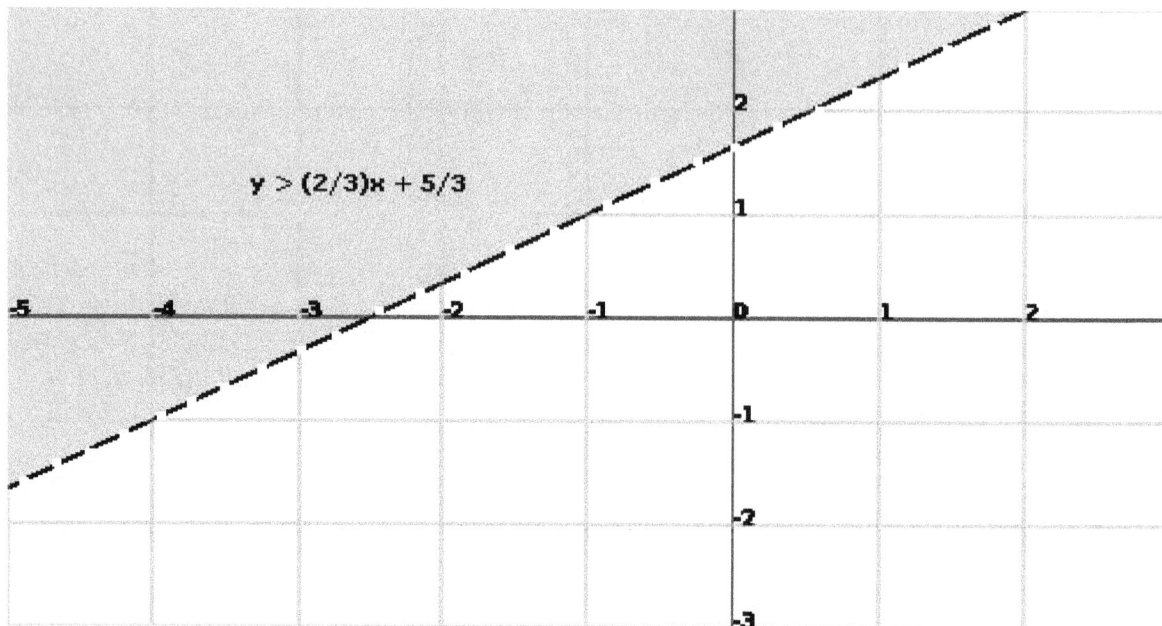

y > (2/3)x + 5/3

2. Let number of compact cars $= x$

Number of sport utility vehicle $= y$

Then

$$11000x + 22000y \leq 1408000$$

$$11x + 22y \leq 1408$$

$$x + 2y \leq 128$$

$$2y \leq -x + 128$$

$$y \leq -0.5x + 64$$

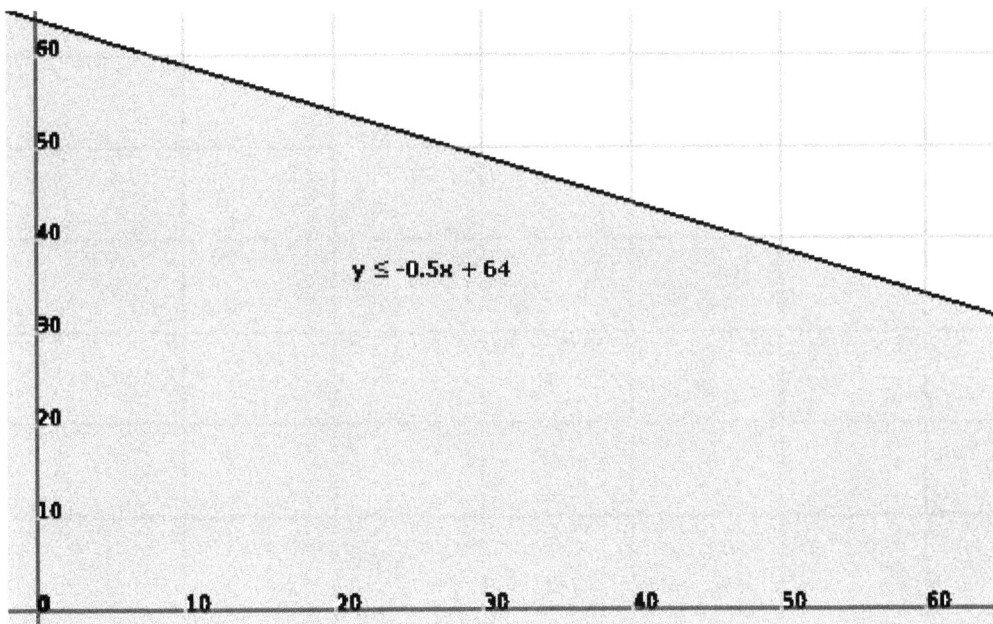

$y \leq -0.5x + 64$

**3.**   Let Math book = $x$

Science book = $y$

Then,

$$8x + 10y \leq 90$$
$$4x + 5y \leq 45$$
$$5y \leq -4x + 45$$
$$y \leq -(4/5)\,x + 9$$

If 7 Math books are bought, Put $x = 7$ to find maximum number of Science books.

$$y \leq -(4/5)\,x + 9$$
$$y \leq -(4/5)(6) + 9$$
$$y \leq -(24/5) + 9$$
$$y \leq 21/9$$
$$y \leq 2.33$$

So Maximum 2 Science books can be bought.

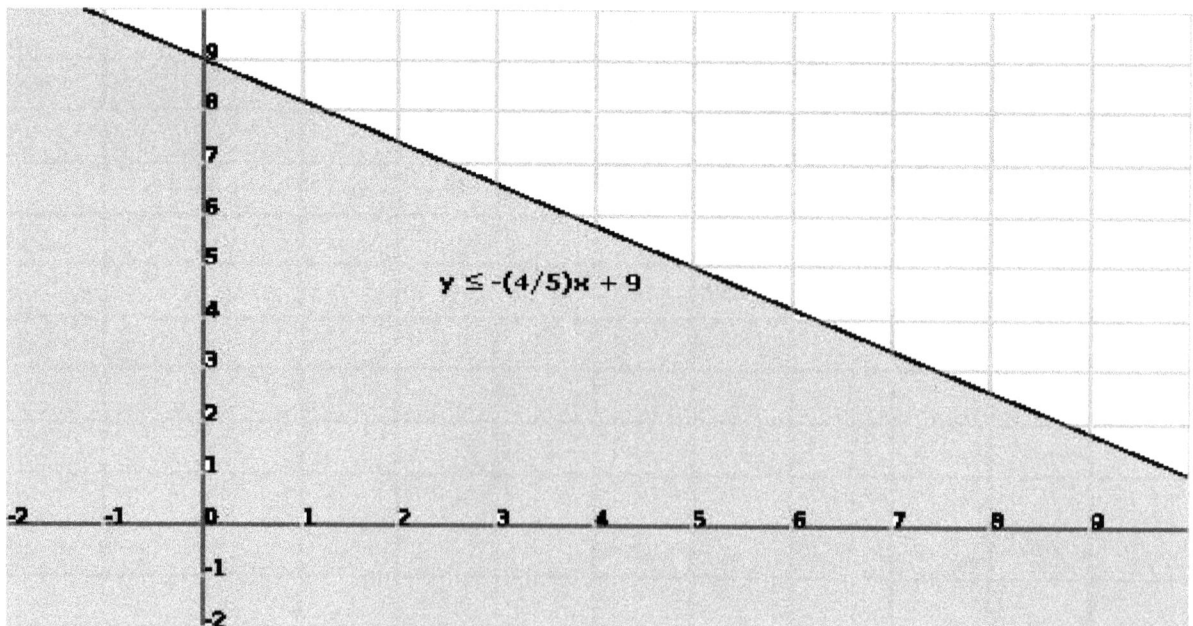

$$y \leq -(4/5)x + 9$$

# Mini-Assessment

1.  Choose the inequality that represents the graph.

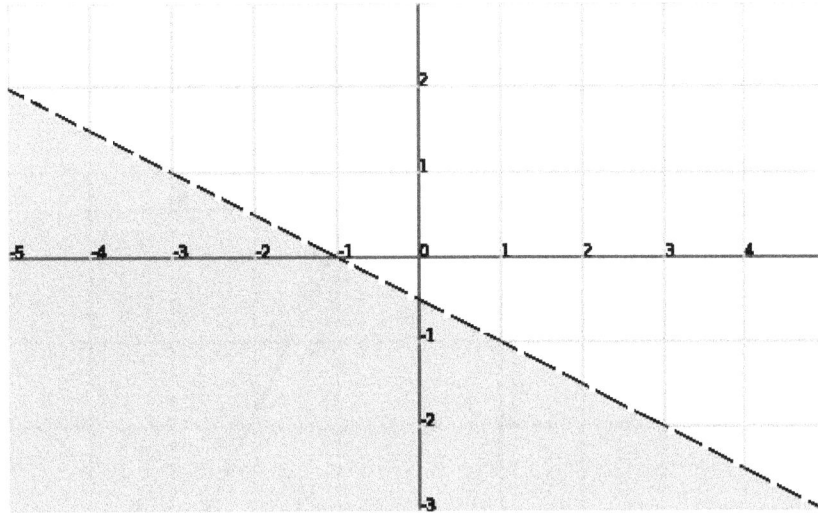

**A.** $y < (1/2)x - 1/2$      **B.** $y < (-1/2)x + 1/2$      **C.** $y < (-1/2)x - 1/2$

**D.** $y < (1/2)x + 1/2$      **E.** $y > (-1/2)x - 1/2$

**2.** Choose the inequality that represents the graph.

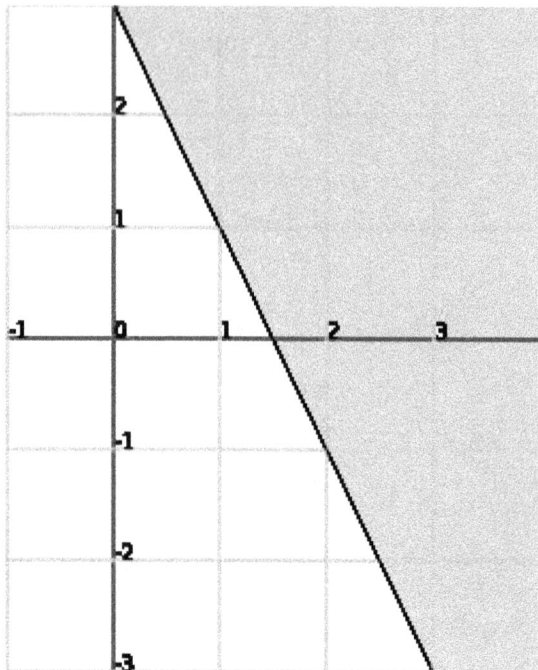

**A.** $2x + y > 3$  **B.** $2x + y \geq 3$  **C.** $-2x + y \geq 3$

**D.** $2x - y \geq 3$  **E.** $2x + y \leq 3$

**3.** Choose the inequality that represents the graph.

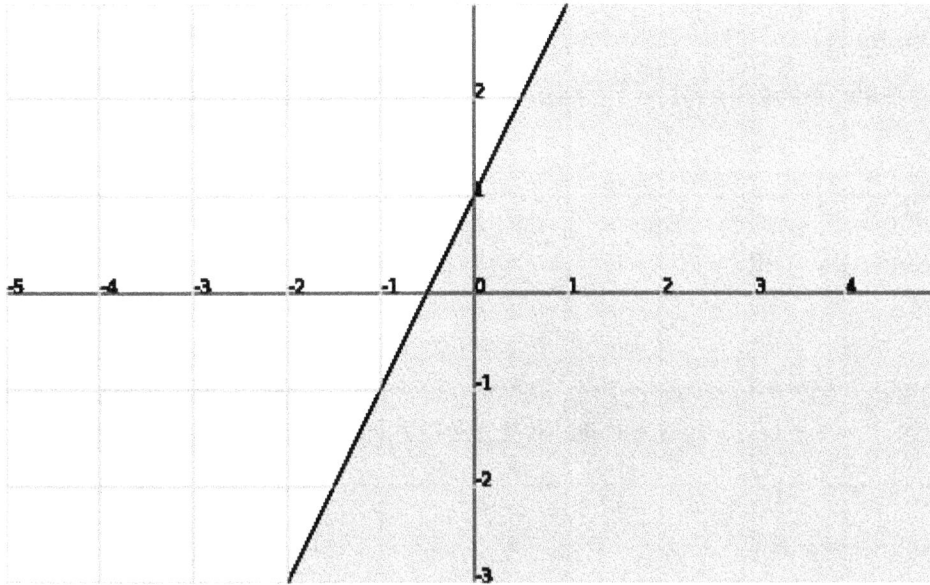

**A.** $y - 2x > 1$          **B.** $y + 2x \leq -1$          **C.** $y - 2x \leq -1$

**D.** $y - 2x \leq 1$          **E.** $y + 2x \leq 1$

**4.** How to decide which side of the line is the solution set of the given inequality.

**5.** The line of the graph is solid, if the inequality is of the form _____ or _____.

**6.** The line of the graph is dotted, if the inequality is of the form _____ or _____.

**7.** A dozen pencils cost 10$ and a dozen pens cost 27$. If you have 100$, write an inequality that shows the possible numbers of pens and pencils to be bought in 100$.

# Mini-Assessment Answers and Explanations

**1.** Ans: Option (c)

Looking at the graph.

$m = -y/x = -(-1/2)/-1 = -1/2$

Y- intercepy $= c = -1/2$

So eq. of the line will be

$y = mx + c$

$y = (-1/2) x - (1/2)$

the line in the graph is dotted and its lower part is shaded, so < will be used in the inequality.

$y < (-1/2) x - (1/2)$

So option (c) is true.

**2.** Ans : Option (b)

Looking at the graph.

$m = -y/x = -3/(3/2) = -2$

Y- intercepy $= c = 3$

So eq. of the line will be

$y = mx + c$

$y = -2x + 3$

the line in the graph is solid and its upper part is shaded, so $\geq$ will be used in the inequality.

$y \geq -2x + 3$

$y + 2x \geq 3$

So option (b) is true.

3. Ans : Option (d)

Looking at the graph.

$m = -y/x = -1/(-1/2) = 2$

$Y$- intercepy $= c = 1$

So eq. of the line will be

$y = mx + c$

$y = 2x + 1$

the line in the graph is solid and its lower part is shaded, so $\leq$ will be used in the inequality.

$y \leq 2x + 1$

$y - 2x \leq 1$

So option (d) is true.

4. Test point (0, 0) is plugged into the inequality, if its true then the side where (0, 0) lies is the solution set and if its false, then other side will be the solution set.

5. $y \leq mx + b, y \geq mx + b$

6. $y < mx + b, y > mx + b$

7. Let a dozen pencils $= x$

a dozen pens $= y$

$10x + 27y \leq 100$

## Lesson Reflection

**(a)** Analyze situations involving linear functions and formulate linear equations or inequalities to solve problems.

**(b)** investigate methods for solving linear equations and inequalities using concrete models, graphs, and the properties of equality, select a method, and solve the equations and inequalities.

**(c)** interpret and determine the reasonableness of solutions to linear equations and inequalities.

# Lesson 3
# Distance and Midpoint Formula

### Lesson Description:

This lesson is designed to help students understand how to use both the distance and midpoint formula when calculating distance and midpoint between two points on a plane. Please be sure to utilize the questions to help spark student engagement and cover the vocabulary that is associated with this specific tutoring session. For your own knowledge, sample responses have been provided to guide you as well.

### Learning Objective(s):

In today's lesson, the learner will calculate distance and midpoint between two points on a plane in 3 out of 4 trials with 75% or above accuracy.

## Introduction

You may have heard of the distance and midpoint formula as it is used to find the distance between two points within the coordinate plane. But, did you know that it could be used to solve other situations. Imagine you are putting up a fence and need to cement the fence posts so that the fence can be upright. As you know, cement takes a couple days to fully set so this will require for additional support to ensure that the cement dries properly. How could the distance and midpoint formula in this situation?

## Questions to Engage Students

➤ If two points are located on a horizontal line in a coordinate plane, how will you find the distance between the points?

➤ If two points are located on a vertical line in a coordinate plane, how will you find the distance between them?

➤ How to find the midpoint of any two numbers?

## Connect Learning Objective(s) Student's Lives

**A)** Distance formula is used when different locations are identified as ordered pairs on the coordinate plane or on the map and we need to find the distance between them.

**B)** Mid point formula is helpful when different locations are identified as ordered pairs on the coordinate plane or on the map and we need to find the midpoint between any two given locations.

# Specific Vocabulary Covered

### Co-ordinate Plane

A two dimensional surface containing '$x$-axis' and '$y$-axis' is called a co-ordinate plane.

### Ordered Pairs

Ordered pair is used to show the location of a point on the co-ordinate plane. It is usually written as $(x, y)$.

### Distance Formula

Distance formula is used to calculate the distance between any two points on the coordinate plane.

### Mid Point Formula

Mid point formula is used to calculate the mid point between any two points on the coordinate plane.

# Direct & Guided Instruction: Modeling For You and Working With You

## Distance Formula

Distance formula is used to calculate the distance between the two points on the coordinate plane. The distance '$d$' between two points and is given by,

$$d = \sqrt{(y_2 - y_1)^2 + (x_2 - x_1)^2}$$

## Mid Point Formula

To calculate a point exactly in the middle of the two given points on the coordinate plane, Mid Point Formula is used.

Given two points $(x_1, y_1)$ and $(x_2, y_2)$, Mid point formula can be given as,

$$\textbf{Mid Point} = (\frac{x_1 + x_2}{2}, \frac{y_1 + y_2}{2})$$

☞ **Problem 1:** **Find the distance between (–1, 4) and (5, 8)?**

### Teacher Questions

1. What are the values for $x_1$ and $y_1$ ?
2. What are the values of $x_2$ and $y_2$ ?
3. Why the distance formula is used?
4. Can you write the distance formula?

### Solution

$d = \sqrt{(y_2 - y_1)^2 + (x_2 - x_1)^2}$

$d = \sqrt{(8 - 4)^2 + (5 - (-1))^2}$

$d = \sqrt{(4)^2 + (5 + 1)^2}$

$d = \sqrt{(4)^2 + (6)^2}$

$d = \sqrt{16 + 36}$

$d = \sqrt{52}$

☞ **Problem 2:** Find the midpoint between (8, 4) and (8,-4)?

## Teacher Questions

1. What are the values for $x_1$ and $y_1$ ?
2. What are the values of $x_2$ and $y_2$ ?
3. Why the midpoint formula is used?
4. Can u write the midpoint formula?

### Solution

$$\text{Mid Point} = \left(\frac{8+8}{2}, \frac{4+(-4)}{2}\right)$$

$$\text{Mid Point} = \left(\frac{16}{2}, \frac{4-4}{2}\right)$$

$$\text{Mid Point} = \left(8, \frac{0}{2}\right)$$

$$\text{Mid Point} = (8, 0)$$

☞ **Problem 1:** A triangle has the vertices $(3, 5)$, $(-1, -1)$ and $(5, -2)$. Find the length of each side of the triangle?

## Teacher Questions

1. If the vertices are labeled as $A$, $B$ and $C$, what will be the three sides?
2. How will you represent the length of sides of the triangle on the coordinate plane?
3. How will you find the length of each side?

**Solution**

Plot the points and use distance formula to calculate $AB$, $BC$ and $CA$.

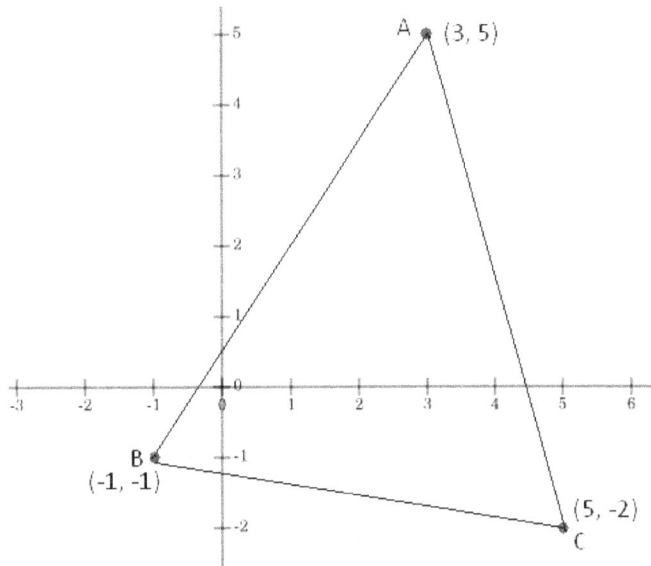

$$AB = \sqrt{(-1-3)^2 + (-1-5)^2}$$

$$AB = \sqrt{(-4)^2 + (-6)^2}$$

$$AB = \sqrt{16+36} = \sqrt{52}$$

$$BC = \sqrt{(-1-5)^2 + (-1+2)^2}$$

$$BC = \sqrt{(-6)^2 + (1)^2}$$

$$BC = \sqrt{36+1} = \sqrt{37}$$

$$CA = \sqrt{(3-5)^2 + (5+2)^2}$$

$$CA = \sqrt{(-2)^2 + (7)^2}$$

$$CA = \sqrt{4+49} = \sqrt{53}$$

☞ <u>**Problem 2:**</u> **On a map's coordinate, location of a player of hockey team *A* is (−1, −3) and location of another player of hockey team *B* is (4, 9). The ball is exactly in the middle of both the players. Find the location of the ball on the plane?**

## Teacher Questions

(i). What does it mean that "The ball is exactly in the middle of both the players"?

(i). How will you find the midpoint?

## Solution

Given that $A(-1, -3)$ and $B(4, 9)$.

Location of ball on the coordinate plane = Midpoint of Points $A$ and $B$

Using Mid point formula

$$\left(\frac{x_1 + x_2}{2}, \frac{y_1 + y_2}{2}\right)$$

$$\left(\frac{-1+5}{2}, \frac{-3+9}{2}\right)$$

$$\left(\frac{4}{2}, \frac{6}{2}\right) = (2,3)$$

# Video Suggestions for Tutoring Sessions

Please conduct a search on either YouTube or Teacher Tube to find appropriate videos for this lesson. Below are some suggested title searches:

➤ Distance in the Coordinate Plane

➤ Mid Point of a Line

➤ Distance Formula

# Independent Instruction: Working on Your Own

## Questions

☞ <u>**Problem 1:**</u> A trapezoid has the vertices (2,4), (3,1), (-4,1) and (–3,–2). Show that the trapezoid is isosceles?

☞ <u>**Problem 2:**</u> Refer to the problem 1 and find the midpoint of each side of the trapezoid.

☞ <u>**Problem 3:**</u> Find the midpoint between the points (8, 7) and (–4, 3). Using distance formula verify that the midpoint you calculated is true?

**1.** Let $A\,(2,4)$, $B(3,1)$, $C(-4,1)$, $D(-3,-2)$.

Using Distance formula, we will now calculate the length of each side of the trapezoid.

$$AB = \sqrt{(3-2)^2 + (1-4)^2} = \sqrt{(1)^2 + (-3)^2} = \sqrt{1+9} = \sqrt{10}$$

$$BC = \sqrt{(-4-3)^2 + (1-1)^2} = \sqrt{(-7)^2 + (0)^2} = \sqrt{49} = 7$$

$$CD = \sqrt{(-3+4)^2 + (-2-1)^2} = \sqrt{(1)^2 + (-3)^2} + \sqrt{1+9} = \sqrt{10}$$

$$DA = \sqrt{(2+3)^2 + (4+2)^2} = \sqrt{(5)^2 + (6)^2} = \sqrt{25+36} = \sqrt{61}$$

We can see that $AB = CD$, therefore, trapezoid $ABCD$ is an isosceles trapezoid.

**2.** $A(2,4)$, $B(3,1)$, $C(-4,1)$, $D(-3,-2)$.

Midpoint of $AB$

Midpoint of $BC$

Midpoint of $CD$

Midpoint of $DA$

$$AB = \left(\frac{2+3}{2}, \frac{4+1}{2}\right) = \left(\frac{5}{2}, \frac{5}{2}\right) = (2.5, 2.5)$$

$$BC = \left(\frac{3-4}{2}, \frac{1-1}{2}\right) = \left(\frac{-1}{2}, \frac{0}{2}\right) = (-0.5, 0)$$

$$CD = \left(\frac{-4+3}{2}, \frac{1+2}{2}\right) = \left(\frac{-1}{2}, \frac{3}{2}\right) = (-0.5, 1.5)$$

$$DA = \left(\frac{2-3}{2}, \frac{4-2}{2}\right) = \left(\frac{-1}{2}, \frac{2}{2}\right) = (-0.5, 1)$$

**3.** Let $A(8,7)$, $B(-4, 3)$ and midpoint be $C$.

Mid Point $C$ between $A$ and $B$ =

$$\left(\frac{8-4}{2}, \frac{7+3}{2}\right)$$

$$\left(\frac{4}{2}, \frac{10}{2}\right) = (2,5)$$

If the midpoint is true then $AC = BC$. To verify we will calculate $AC$ and $BC$ using distance formula.

$$AC = \sqrt{(8-2)^2 + (7-5)^2} = \sqrt{(6)^2 + (2)^2} = \sqrt{36+4} = \sqrt{40}$$

$$BC = \sqrt{(-4-2)^2 + (3-5)^2} = \sqrt{(-6)^2 + (-2)^2} = \sqrt{36+4} = \sqrt{40}$$

Since, $AC = BC$, therefore, the midpoint calculated is true.

# Mini-Assessment

**1.** What is the midpoint for the line segment HK?

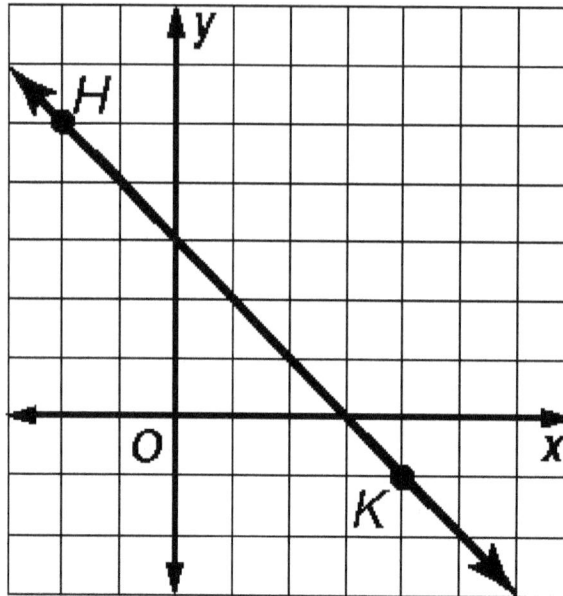

| | | |
|---|---|---|
| **A.** $(1, 2)$ | **B.** $(3, 3)$ | **C.** $(-3, -3)$ |
| **D.** $(-1, 2)$ | **E.** $(-1, -2)$ | |

**2.** What is the distance between $(5, 1)$ and $(-4, -2)$?

| | | |
|---|---|---|
| **A.** $45$ | **B.** $10\sqrt{3}$ | **C.** $3\sqrt{10}$ |
| **D.** $9\sqrt{10}$ | **E.** $10\sqrt{9}$ | |

**3.** The endpoints of a line $AB$ are $(-1, 4)$ and $(2, -7)$. What are the coordinates of the mid-point of line $AB$?

**A.** $(0.5, 1.5)$      **B.** $(1.5, 0.5)$      **C.** $(-0.5, 1.5)$

**D.** $(0.5, -1.5)$      **E.** $(1.5, -0.5)$

**4.** Refer to the graph and find the length of the line segment $WZ$?

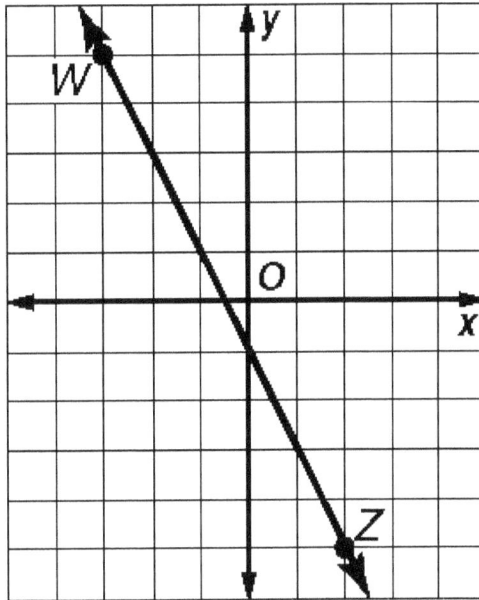

**5.** On a map Lisa's Home is located at $(6, 8)$ and Park is located at $(-10, 4)$. If Lisa's home is exactly in the middle of the Hospital and the Park. Find the location of the hospital?

**6.** The center of the circle is located at $(1, 2)$. If one end point of the diameter of the circle is at $(5, 4)$, find the length of the radius?

**7.** Three points $(3, 2)$, $(-1, 0)$ and $(0, -5)$ form a triangle. Find the midpoint of each side of the triangle?

# Mini-Assessment Answers and Explanations

**1.** A

**2.** C

**3.** D

**4.** $W(-3, 5)$ and $Z(2, -5)$

$WZ = \text{sqrt } [(-3-2)^2+(5+5)^2]$

$WZ = \text{sqrt } [(-5)^2+(10)^2]$

$WZ = \text{sqrt } [25+100]$

$WZ = \text{sqrt } [125]$

$WZ = 5 \text{ sqrt } (5)$

## Answer 5:

Let $x, y$ be the coordinates of the hospital.

$(6, 8) = ((x - 10)/2, (y + 4)/2)$ --- (i)

Equating '$x$' coordinates of eq. (i)

$(x - 10)/2 = 6$

$X - 10 = 12$

$x = 12 + 10 = 22$

Equating '$y$' coordinate of eq. (i):

$(y + 4)/2 = 8$

$y + 4 = 16$

$y = 16 - 4 = 12$

Therefore coordinates of the hospital are $(22, 12)$

**Answer 6:**

Radius = Distance between the center of the circle and one end of the diameter.

Radius = sqrt [(5-1)^2 + (4-2)^2]

Radius = sqrt [(4)^2 + (2)^2]

Radius = sqrt [16 + 4]

Radius = sqrt (20)

Radius = 2 sqrt (5)

**Answer 7:**

Let $A$ (3, 2), $B$ (–1, 0) and $C$ (0, –5)

Midpoint of side AB = ((3 – 1)/2, (2 + 0)/2) = (2/2, 2/2) = (1, 1)

Midpoint of side BC = ((–1 + 0)/2, (0 – 5)/2) = (–1/2, –5/2)

Midpoint of side CA = ((3 + 0)/2, (2 – 5)/2) = (3/2, –3/2)

## Lesson Reflection

### Distance Formula

Distance formula is used to calculate the distance between the two points on the coordinate plane.

The distance '$d$' between two points $x_1$, $y_1$ and $(x_2, y_2)$ is given by,

$$d = \sqrt{(x_2 - x_1)^2 + (y_2 - y_1)^2}$$

### Mid Point Formula

To calculate a point exactly in the middle of the two given points on the coordinate plane, Mid Point Formula is used.

Given two points $(x_1, y_1)$ and $(x_2, y_2)$ , Mid point formula can be given as,

$$\text{Mid Point} = \left(\frac{x_1 + x_2}{2}, \frac{y_1 + y_2}{2}\right)$$

# Lesson 4
# Slope of a Line

## Lesson Description:

This lesson is designed to help students find the slope of a line. Additionally, learners will have an opportunity to find the slope of both parallel and perpendicular lines. Please be sure to utilize the questions to help spark student engagement and cover the vocabulary that is associated with this specific tutoring session. For your own knowledge, sample responses have been provided to guide you as well.

## Learning Objective(s):

In today's lesson, the learner will find the slope of a line, including parallel and perpendicular lines with 3 out of 4 trials with 75% or above accuracy

## Introduction

What comes to mind as you think about a slope of a line? When you think about skiing, how are slopes used? How do you think about slopes when building roads? Slopes are important to help determine the steepness. Even when it comes to building stairs, one must consider the slope to ensure that it is not too steep to walk on. As you can see, the slope of a line is a very important skill to learn more about as it is used to measure steepness.

## Questions to Engage Students

➤ What do you understand by the word 'slope'?

➤ Can you give some examples of slope?

## Connect Learning Objective(s) Student's Lives

**A.** Slope is used while building roof pitch, road grades, wheel chair ramps etc. Other examples are,

**B.** Mountain boarding,

**C.** Roller coasters,

**D.** Avalanches,

# Specific Vocabulary Covered

## Slope of a Line

Slope of a line is the measure of steepness of a line. Slope remains constant for the whole line.

## Undefined Slope

A vertical line has an undefined slope.

## Zero Slope

A horizontal line has a zero slope.

## Parallel Lines

Two never intersecting lines are called parallel lines.

## Perpendicular Lines

Two lines making an angle of 90° with each other are called perpendicular lines.

# Direct & Guided Instruction: Modeling For You and Working With You

## Calculating Slope of Line

1. When two points of the line are given

    If two points $(x_1, y_1)$ and $(x_2, y_2)$ are the points of a straight line then slope of the line is calculated as, $Slope = \dfrac{y_2 - y_1}{x_2 - x_1}$

2. When graph of the line is given

    Consider any two points on the line.

    Draw a horizontal line (Run) from point and a vertical line (Rise) from the other point.

    Slope = Rise/Run

    In the figure, slopes of two different lines are shown.

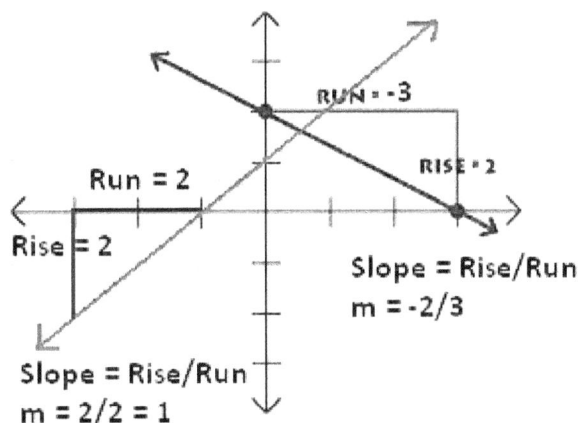

RUN = -3

RISE = 2

Run = 2

Rise = 2

Slope = Rise/Run
m = -2/3

Slope = Rise/Run
m = 2/2 = 1

## Slopes of Parallel Lines

If two non vertical lines are parallel to each other, they will have same slope.

$$If$$
$$l_1 \parallel l_2$$
$$then$$
$$m_1 = m_2$$

## Slopes of Perpendicular Lines

If two lines are perpendicular to each other, then the product of their slopes is '-1'. In other words,

$$If$$
$$l_1 \perp l_2$$
$$then$$
$$m_1.m_2 = -1$$
$$m_1 = \frac{-1}{m_2}$$

☞ **Problem 1:** **Find the slope of a line passing through the points (–1, 0) and (3, 5)?**

### Teacher's Question

1. What are the values of $x_1$ and $x_2$ ?

2. What are the values of $y_1$ and $y_2$ v ?

3. How will define the slope of a line in your own words?

### Solution

Slope of a line passing through the points (–1, 0) and (3, 5),

$$Slope = \frac{y_2 - y_1}{x_2 - x_1}$$
$$Slope = \frac{5 - 0}{3 - (-1)}$$
$$Slope = \frac{5}{3+1} = \frac{5}{4}$$

☞ **Problem 2:** **Line 'm' is perpendicular to the line 'l' passing through the points (4, 6) and (–2, 4). Find the slope of the line 'm' ?**

1. How will you calculate the slope of line 'l' ?

2. What is the relation between the slopes of two perpendicular lines?

3. Will this relation be helpful in finding the slope of line 'm' ? How?

**Solution**

Let $m_1$ be the slope of line 'l' passing through the points (4, 6) and (–2, 4), then,

$$m_1 = \frac{y_2 - y_1}{x_2 - x_1}$$

$$m_1 = \frac{4 - 6}{-2 - 4}$$

$$m_1 = \frac{-2}{-6} = \frac{1}{3}$$

Let $m_2$ be the slope of the line 'm' .

As, lines 'l' and 'm' are perpendicular, therefore,

$$m_1 . m_2 = -1$$

$$\left(\frac{1}{3}\right) . m_2 = -1$$

$$m_2 = -3$$

☞ **Problem 1:** **One line is passing through the points (3, –2) and (5, –3), where as the other line is passing through the points (4, 7) and (3, 5). Find whether the line are parallel, perpendicular or neither?**

## Teacher Questions

1. How to find the slope of a line when two points are given?

2. If slope of two lines are known, how will you get to know that the lines are parallel?

3. If slope of two lines are known, how will you get to know that the lines are perpendicular?

## Solution

Slope of the first line,

$$m_1 = \frac{-3-(-2)}{5-3} = \frac{-3+2}{2} = \frac{-1}{2}$$

Slope of the second line,

$$m_2 = \frac{5-7}{3-4} = \frac{-2}{-1} = 2$$

Slopes are not equal, therefore, lines are not parallel. Let's check what their product comes out to be.

$$m_1.m_2 = \left(\frac{-1}{2}\right)(2) = -1$$

Therefore, lines are perpendicular to each other.

☞ **Problem 2**: Find the slope of the line from the graph?

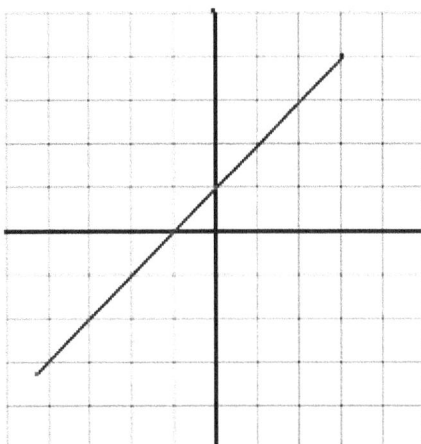

**Teacher Questions**

1. How slope of a line is calculated?
2. Locate any two points on the graph?
3. How will you calculate the 'RISE'?
4. How to calculate the 'RUN'?

**Solution**

Mark any two points on the line.
Draw a horizontal line from one point
To calculate 'RUN' and a vertical line
from the other point to calculate 'RISE'.
These horizontal and vertical lines
should be drawn in a way that both
intersect each other as shown in the figure.

RISE = 4

RUN = 4

Slope = RISE/RUN = 4/4 = 1

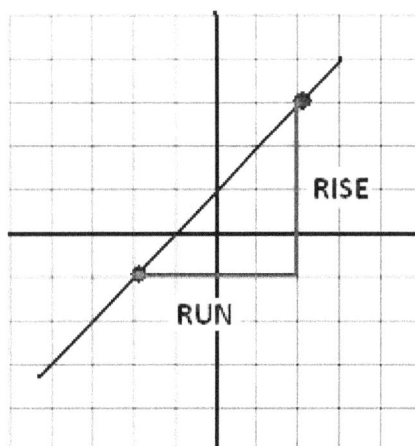

# Video Suggestions for Tutoring Sessions

Please conduct a search on either YouTube or Teacher Tube to find appropriate videos for this lesson. Below are some suggested title searches:

➤ Finding the Slope of a line

➤ Determining the Slope of a Line from the Graph

# Independent Instruction: Working on Your Own

## Questions

☞ <u>Problem 1:</u>  Find the slope of a line perpendicular to the line passing through the points (2, 10) and (8, 7)?

☞ <u>Problem 2:</u>  Find the slope of the line shown in the graph.

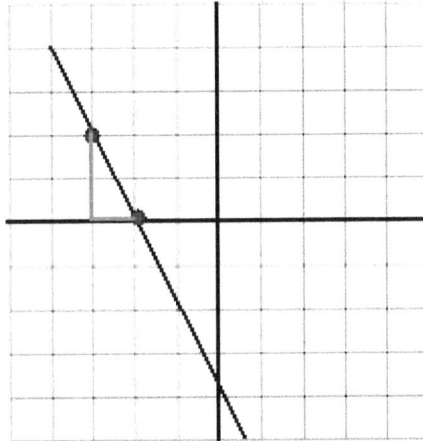

☞ <u>Problem 3:</u>  The slope of a line is 7/3 and the line is passing through the points (5, –2) and (8, $p$). Find the value of $p$?

**Solution**

1. Slope of a line passing through the points (2, 10) and (8, 7),

$$m_1 = \frac{7-10}{8-2} = \frac{-3}{6} = \frac{-1}{2}$$

Let $m_2$ be the slope of the line perpendicular to the line with slope $m_1$, then,

$$m_1 . m_2 = -1$$

$$\left(\frac{-1}{2}\right) \cdot m_2 = -1$$

$$m_2 = 2$$

Therefore, slope of the line perpendicular to the given line is 2.

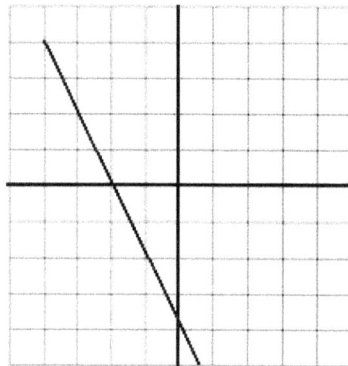

2. Locate two points on the graph to find RISE and RUN.

RISE = –2

RUN = 1

Therefore,

Slope = RISE/RUN

Slope = –2/1 = –2

3. $Slope = \dfrac{y_2 - y_1}{x_2 - x_1}$

$$\frac{7}{3} = \frac{p-(-2)}{8-5}$$

$$\frac{7}{3} = \frac{p+2}{3}$$

$$7 = p+2$$

$$p = 7-2 = 5$$

# Mini-Assessment

☞ **Problem 1:**  **A horizontal line has a slope**

**A.** 0          **B.** 1          **C.** −1          **D.** Undefined


☞ **Problem 2:**  **Two lines are said to be perpendicular if**

**A.** Sum of their slopes is 0.

**B.** Product of their slopes is 1.

**C.** Sum of their slopes is −1.

**D.** Product of their slopes is −1.


☞ **Problem 3:**  **Find the slope of a line that contains the two points.**

$$(-2, 8) \text{ and } (1, 2)$$

**A.** ½          **B.** −2          **C.** −½          **D.** 2


☞ **Problem 4:**  **Find the slope of the line perpendicular to the line in the graph?**

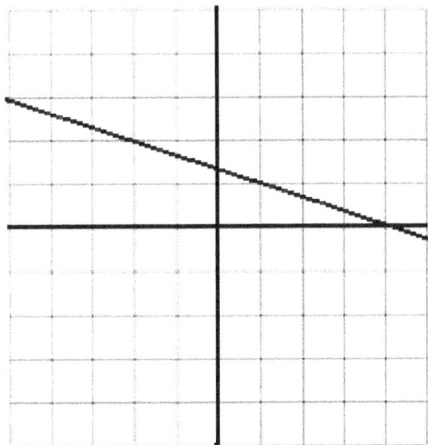

☞ <u>Problem 5</u>: Line '*p*' passes through the points (0, 0) and (5, 6) and line '*q*' passes through the points (−3, −2) and (−4, −3). Find that the line '*p*' and '*q*' are parallel, perpendicular or neither?

☞ <u>Problem 6</u>: The slope of a line is −3/2 passing through the points (−2, 3) and (*a*, 0). Find *a*?

☞ <u>Problem 7</u>: Find the slope of a line parallel to the line passing through the points (8, −5) and (−10, 7)?

# Mini-Assessment Answers and Explanations

**1.** A

**2.** C

**3.** D

**4.** Identify to points on the graph $(-2, 2)$ and $(1, 1)$ and calculate RISE and RUN.

RISE = $-1$

RUN = 3

Slope = RISE/RUN = $-1/3$

Slope of the line perpendicular to the given line = $-1/$(slope of the given line) = 3

**5.** Slope of line '$p$' = $(6 - 0)/(5 - 0) = 6/5$

Slope of line '$q$' = $(-3 + 2)/(-4 + 3) = -1/-1 = 1$

Slope of both lines are not equal, so they are not parallel.
Neither the product of both slopes is $-1$.

So the lines are neither parallel nor perpendicular.

**6.** Slope = (Change in $y$-values)/(Change in $x$-values)

$-3/2 = (0 - 3)/(a + 2)$

$-3/2 = -3/(a + 2)$

$-3(a + 2) = -3(2)$

$a + 2 = 2$

$a = 2 - 2 = 0$

**7.** Slope = (Change in $y$-values)/(Change in $x$-values)

Slope = $(7 + 5)/(-10 - 8)$

Slope = $12/-18$

Slope = $2/(-3) = -2/3$

Slope of the line parallel to the given line = $-2/3$

## Lesson Reflection

### Slope of a Line

Slope of a line is the measure of steepness of a line. Slope remains constant for the whole line.

Slope of line = RISE/RUN

= (Change in $y$-values)/ (Change in $x$-values)

### Slopes of Parallel Lines

If two non vertical lines are parallel to each other, they will have same slopes.

### Slopes of Perpendicular Lines

If two lines are perpendicular to each other, then the product of their slopes is '–1'.

# Lesson 5
# Equations of Lines

## Lesson Description:

This lesson is designed to help students write equations of lines. Please be sure to utilize the questions to help spark student engagement and cover the vocabulary that is associated with this specific tutoring session. For your own knowledge, sample responses have been provided to guide you as well.

## Learning Objective(s):

In today's lesson, the learner will find the equations of lines in 3 out of 4 trials with 75% or above accuracy.

# Introduction

Have you ever had to take a cab? What did you learn about the cost of a cab drive from the airport to your hotel? Did you know that the cost of the cab drive could be determined by using the equation of a line to solve for it? Equations of lines are used to find a variety of relationships in mathematics. These relationships include distance-time, currency, weight, conversions of temperatures, and sometimes in business.

## Questions to Engage Students

**1.** If a graphical representation of a line is given, how will you find the slope of the line?

**2.** Is there any way to find the slope of the line if any two points on the line are provided?

**3.** Do you know the slope of a horizontal and a vertical line?

**4.** Can you write equation of a line using two variables 'x' and 'y'?

## Connect Learning Objective(s) Student's Lives

Linear equations are widely used in real life applications.

**A.** Linear equations are used in Economics. Supply and Demand is an example of a linear relationship.

**B.** In Mathematics, e.g Distance-Time relationship, Calculating travel times, Conversions of units of temperatures, currency, Weight etc.

**C.** Used in Business to determine prices, to create plans like cell phone plans etc and to derive values.

# Specific Vocabulary Covered

## Slope of a line

Slope is the measure of steepness of a line.

Slope = Rise/Run = (Change in $y$)/(Change in $x$)

## Equation of Horizontal line

The line is horizontal when slope = 0. Equation of horizontal line is always $y = b$.

## Equation of Vertical Line

The line is vertical when slope is Undefined. Equation of horizontal line is always $x = a$.

## Y-intercept

Y-intercept is where the line crosses the $y$-axis. It is denoted by '$b$'.

# Direct & Guided Instruction: Modeling For You and Working With You

## Ways to Write Equation of Line

1. Standard form

   $Ax + By = C$

2. Slope Intercept Form

   $y = mx + b$

3. Point Slope Form

   $y - y_1 = m(x - x_1)$

☞ **Problem 1: Find Equation of a line passing through the point (–1, 4) having slope –1/2? Write the equation in standard form?**

### Teacher Questions

1. What method will be used to find the equation of line?

2. Write the formula to find the equation of line?

3. Do you know the proper way to write an equation in standard form?

### Solution

$$y - y_1 = m(x - x_1)$$

$$y - 4 = \left(\frac{-1}{2}\right)(x - (-1))$$

$$y - 4 = \left(\frac{-1}{2}\right)(x + 1)$$

$$2(y - 4) = (-1)(x + 1)$$

$$2y - 8 = -x - 1$$

$$x + 2y = 8 - 1$$

$$x + 2y = 7$$

☞ **Problem 2:** A line passes through two points (–3, 2) and (6, –1). Find equation of the line and write it in slope intercept form?

## Teacher Questions

1. Which method you use to find equation of a line when two points of the line are given?

2. Write the formula to find the equation of a line using two points?

3. Is there any other way to find equation of a line, when two points are given?

4. What is the slope intercept form?

## Solution

**Slope**

$$m = \frac{y_2 - y_1}{x_2 - x_1} = \frac{-1-2}{6+3} = \frac{-3}{9} = \frac{-1}{3}$$

**Equation of the line**

$$y - y_1 = m(x - x_1)$$

$$y - 2 = \frac{-1}{3}(x - (-3))$$

$$y - 2 = -\frac{1}{3}(x + 3)$$

$$y - 2 = -\frac{1}{3}x - 1$$

$$y = -\frac{1}{3}x - 1 + 2$$

$$y = -\frac{1}{3}x + 1$$

☞ **Problem 1:** Find the equation of a line passing through the point (2, 1) and perpendicular to the line passing through (1, 4) and (–3, –3)? Write the equation in standard form?

## Teacher Questions

1. Is there any way to visualize the situation?

2. How to find the slope of the required line?

3. Which method will you use to find the equation of the line?

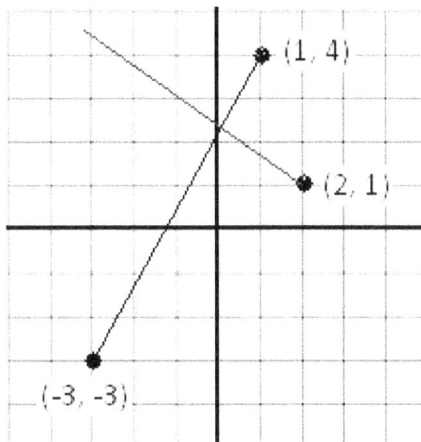

---

**Solution**

Slope of the line passing through the points (1, 4) and (–3,–3),

$$m = \frac{-3-4}{-3-1} = \frac{-7}{-4} = \frac{7}{4}$$

Slope of the required line perpendicular to the line with slope $m = m_1 = \frac{-4}{7}$

Equation of the line passing through the point (2, 1) will be,

$$y - y_1 = m_1(x - x_1)$$

$$y - 1 = \frac{-4}{7}(x - 2)$$

$$7y - 7 = -4x + 8$$

$$4x + 7y = 8 + 7$$

$$4x + 7y = 15$$

☞ **Problem 2:** **Find the equation of a line passing through the point (1, -1) and parallel to the line passing through the points (1, 4) and (-2,-3)? Write the equation in slope-intercept form?**

## Teacher Questions

1. Is there any way to visualize the lines to get the concept clearer?

2. How will you find the slope of the required line?

3. What is slope-intercept form?

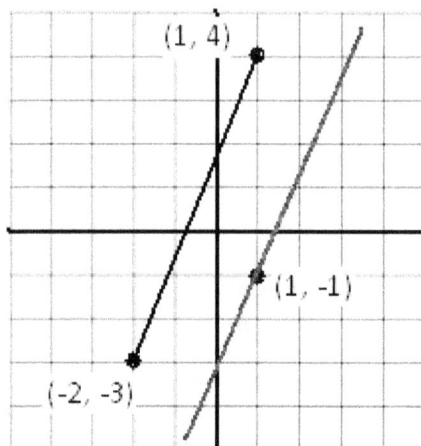

## Solution

Slope of the line passing through the points $(-2, -3)$ and $(1, 4)$,

$$m = \frac{4+3}{1+2} = \frac{7}{3}$$

This line is parallel to the required line passing through $(1, -1)$, so,

Slope of the required line $m_1 = \frac{7}{3}$

Equation of the line passing through the point $(1, -1)$ is,

$$y - y_1 = m_1(x - x_1)$$

$$y - (-1) = \frac{7}{3}(x - 1)$$

$$y + 1 = \frac{7}{3}x - \frac{7}{3}$$

$$y = \frac{7}{3}x - \frac{7}{3} - 1 \Rightarrow y = \frac{7}{3}x - \frac{10}{3}$$

# Video Suggestions for Tutoring Sessions

Please conduct a search on either YouTube or Teacher Tube to find appropriate videos for this lesson. Below are some suggested title searches:

➤ Determine the Equation of a Line Using Slope and y-intercept

➤ Writing an Equation of a Line

➤ Finding the Equation of a Line

➤ Writing the Equation of a Perpendicular Line

➤ Finding the Equation of a Parallel Line

# Independent Instruction: Working On Your Own

## Questions

☞ <u>Problem 1:</u>  Write equation of a line with slope –2 and *y*-intercept –3?

☞ <u>Problem 2:</u>  Find equation of a line passing through the points (–2, –5) and (3, 10)?

☞ <u>Problem 3:</u>  Find equation of a line through the point (–1, 8) and perpendicular to the line $3x + 4y = 1$?

Solution

1.  $m = -2$

    $y$-intercept $= b = -3$

    Equation of line will be,

    $$y = mx + b$$
    $$y = (-2)x + (-3)$$
    $$y = -2x - 3$$

2.  Slope of the line $= m = \dfrac{10+5}{3+2} = \dfrac{15}{5} = 3$

    Equation of the line,

    $$y - y_1 = m(x - x_1)$$
    $$y - 10 = 3(x - 3)$$
    $$y - 10 = 3x - 9$$
    $$y = 3x - 9 + 10$$
    $$y = 3x + 1$$

3.  Given equation of line is $3x + 4y = 1$

    We will write it in slope-intercept form to find the slope,

    $4y = -3x + 1$

    $y = (-3/4)x + (¼)$

    Therefore, Slope of the given line $= -3/4$

    Slope of the line perpendicular to the given line is,

    $m = 4/3$

    Equation of the line perpendicular to the given line through $(-1, 8)$,

    $$y - y_1 = m(x - x_1)$$
    $$y - 8 = \frac{4}{3}(x + 1)$$
    $$3(y - 8) = 4(x + 1)$$
    $$3y - 24 = 4x + 4$$
    $$3y - 4x = 28$$

# Mini-Assessment

☞ <u>**Problem 1:**</u>  **Find equation of a line passing through the point (–2, 1) with slope undefined?**

**A.**  $y = 1$          **B.**  $y = $ Undefined          **C.**  $x = -2$          **D.**  $x = $ Undefined

**E.**  No Equation Exists.


☞ <u>**Problem 2:**</u>  **Find equation of a line passing through the point (3, –7) with zero slope?**

**A.**  $x = 3$          **B.**  $y = 0$          **C.**  $y = -7$          **D.**  $x = 0$

**E.**  No Equation exists.


☞ <u>**Problem 3:**</u>  **Find equation of a line with slope ½ passing through the point (1, 0)?**

**A.**  $2y = x + 2$        **B.**  $x = 2y + 1$        **C.**  $y = 2x - 1$        **D.**  $y = 2x + 1$

**E.**  $x = 2y - 1$


☞ <u>**Problem 4:**</u>  **Find equation of a line with slope –3/5 and *y*-intercept 7 and write in standard form?**


☞ **Problem 5:  Find equation of a line passing through the points (–3, –1) and (0, –2)?**


☞ **Problem 6:  Find equation of a line passing through the point (1, 2) and perpendicular to the line passing through the points (1, 5) and (–2, –4)?**


☞ **Problem 7: Find equation of a line parallel to the line $2x + y = 1$ and passing through the point (4,5)?**

# Mini-Assessment Answers and Explanations

1. C

2. C

3. B

4. $m = -3/5$ and $b = 7$
   Equation of line
   $y = mx + b$
   $y = (-3/5) x + 7$
   $5y = -3x + 35$
   $3x + 5y = 35$

5. Slope $= m = (-2 + 1)/(0 + 3) = -1/3$
   Equation of the line,
   $(y + 2) = (-1/3) (x - 0)$
   $(y + 2) = (-1/3)(x)$
   $3(y + 2) = -x$
   $3y + 6 = -x$
   $x + 3y = -6$

6. Slope of a line passing through the points $(1, 5)$ and $(-2, -4)$,
   Slope $= (-4 - 5)/(-2 - 1) = -9/-3 = 3$
   Slope of the perpendicular to the above line,
   $m = -1/3$
   Equation of the line passing through the point $(1, 2)$ having slope $-1/3$ is,
   $(y - 2) = (-1/3)(x - 1)$
   $3(y - 2) = -(x - 1)$
   $3y - 6 = -x + 1$
   $x + 3y = 1 + 6$
   $x + 3y = 7$

**7.** $2x + y = 1$

Write equation in slope-intercept form to find slope,

$y = -2x + 1$

Therefore, slope $= -2$

As the required line is parallel to the given line, hence their slope are equal.

So, $m = -2$

Equation of line passing through the point $(4, 5)$ with slope $-2$ will be,

$(y - 5) = -2(x - 4)$

$(y - 5) = -2x + 8$

$2x + y = 8 + 5$

$2x + y = 13$

### Equation of Horizontal line

The line is horizontal when slope = 0. Equation of horizontal line is always $y = b$.

### Equation of Vertical Line

The line is vertical when slope is Undefined. Equation of horizontal line is always $x = a$.

### Ways to Write Equation of Line

1. Standard form

   $ax + by = c$

2. Slope Intercept Form

   $y = mx + b$ ($b$ is known as $y$-intercept)

3. Point Slope Form

   $y - y_1 = m(x - x_1)$

# Lesson 6
# Linear Equations & Solutions Using Graphing and Substitution

## Lesson Description:

This lesson is designed to help students solve linear equations by graphing and using both substitution. Please be sure to utilize the questions to help spark student engagement and cover the vocabulary that is associated with this specific tutoring session. For your own knowledge, sample responses have been provided to guide you as well.

## Learning Objective(s):

In today's lesson, the learner will solve linear equations by graphing and using both substitution in 3 out of 4 trials with 75% or above accuracy.

## Introduction

Linear Equations and Solutions are used to solve problems related to cost, interest, and principal amount. People are buy cars with high interest rates tend to pay more over the life of the car loan. Individuals can do their own math if they are familiar with how to solve problems using linear equations and solutions.

## Questions to Engage Students

How will you write a linear equation for the following situations:

**1.** 50 less than 3 times a number is 72.

**2.** Equation to find the cost per orange if 5 oranges cost $12?

**3.** 5 pencils and 6 erasers cost $15?

## Connect Learning Objective(s) Student's Lives

**A)** Linear equations are used to solve the problems related to Cost, Interest and Principal amount.

**B)** Distance, Rate or Time Problems are solved using system of linear equations.

**C)** System of linear equations is also used to solve geometric Applications, e.g finding dimensions of some geometric figure when perimeter and relation between the dimensions is given.

# Specific Vocabulary Covered

### Linear Equation

A linear equations is of the form y=mx+b and it represents a straight line.

### Substitution Method

Substitution Method is a method of solving system of equations, in which one variable is solved in terms of other.

### Graphical Method

Graphical Method is the easiest one. Just graph the linear equations and their point of intersection will be the solution set.

### Independent System

If the system of linear equations has one solution then it is called independent system.

### Inconsistent System

If the system of linear equations has no solution, then the system is known as inconsistent system.

### Dependent System

If the system of linear equations has infinitely many solutions, then this system is called a dependent system.

# Direct & Guided Instruction: Modeling For You and Working With You

## Solving Linear Equations Using Substitution Method

Substitution Method is useful, when in an equation, coefficient of a variable is 1.

So that value of this variable can easily be calculated in terms of other variable.

For example,

$$4x + y = 7$$
$$y = 7 - 4x$$

## Steps to Solve Linear Equations Using Substitution Method

1.  Choose any one equation and calculate the value of one variable, let's say '$x$' in terms of '$y$'.

2.  Substitute the value of '$x$' in other equation.

3.  Solve this equation for the value of '$y$'.

4.  Replace back the value of '$y$' in the first equation to get the value of '$x$'.

☞ **Problem 1:** **Solve the following linear equations using Substitution Method.**

$$5x + 3y = -5$$
$$x + 2y = 6$$

### Teacher Questions

1.  Which equation can easily be used to calculate the value of one variable?

2.  How will you use the value of '$x$'?

3.  How will you calculate the value of '$x$'?

☞ **Problem 1:** **Solve the following linear equations using Substitution Method.**

$5x + 3y = -5$ --------- (i)

$x + 2y = 6$ ------------ (ii)

**Solution**

Solving eq. (ii) for "$x$":

$x = 6 - 2y$ ------------ (iii)

Substitute value of "$x$" in eq. (i),

$5x + 3y = -5$

$5(6 - 2y) + 3y = -5$

$30 - 10y + 3y = -5$

$30 - 7y = -5$

$-7y = -35$

$y = 5$

Substitute Value of "$x$" in eq. (iii),

$x = 6 - 2y$

$x = 6 - 2(5) = 6 - 10 = -4$

So, Solution set is $(x, y) = (-4, 5)$

## Solving Linear Equations Using Graphs

Graph both the equations one the plane. The point of intersection of both the lines will be the solution set.

### REMEMBER!

➤ If the system of linear equations has one intersection point, it has one solution and system is called <u>independent system</u>.

➤ If the system of linear equations has one line, it has infinitely many solutions existing on the whole line and this system is called a <u>dependent system</u>.

➤ If the system of linear equations has no intersection point, then it has no solution and system is known as <u>inconsistent system</u>.

## Steps to Graph a Linear Equation

1. Write the equation in the slope-intercept form i.e. $y = mx + c$.

2. Find $y$-intercept by putting $x = 0$ in the equation and plot it on the plane.

**3.** As $m = y/x$, therefore, starting from the $y$ intercept,

    **i)** Count '$y$' points upward if '$y$' is positive and downward if '$y$' is negative.

    **ii)** Count '$x$' points right if '$x$' is positive and left if '$x$' is negative.

**4.** Plot the point obtained.

**5.** Join both the points to get the graph of the linear equation.

☞ **Problem 2:** **Solve the following system of linear equations graphically.**

$$2x + y = 1$$
$$-x + 2y = 3$$

### Teacher Questions

**1.** How will you write $2x + y = 1$ in slope-intercept form?

**2.** What will be the $y$-intercept of above equation?

**3.** Find the slope of the above equation?

**4.** Write $-x + 2y = 3$ in slope-intercept form?

**5.** What is the $y$-intercept for the above equation?

**6.** Calculate the slope for the above equation?

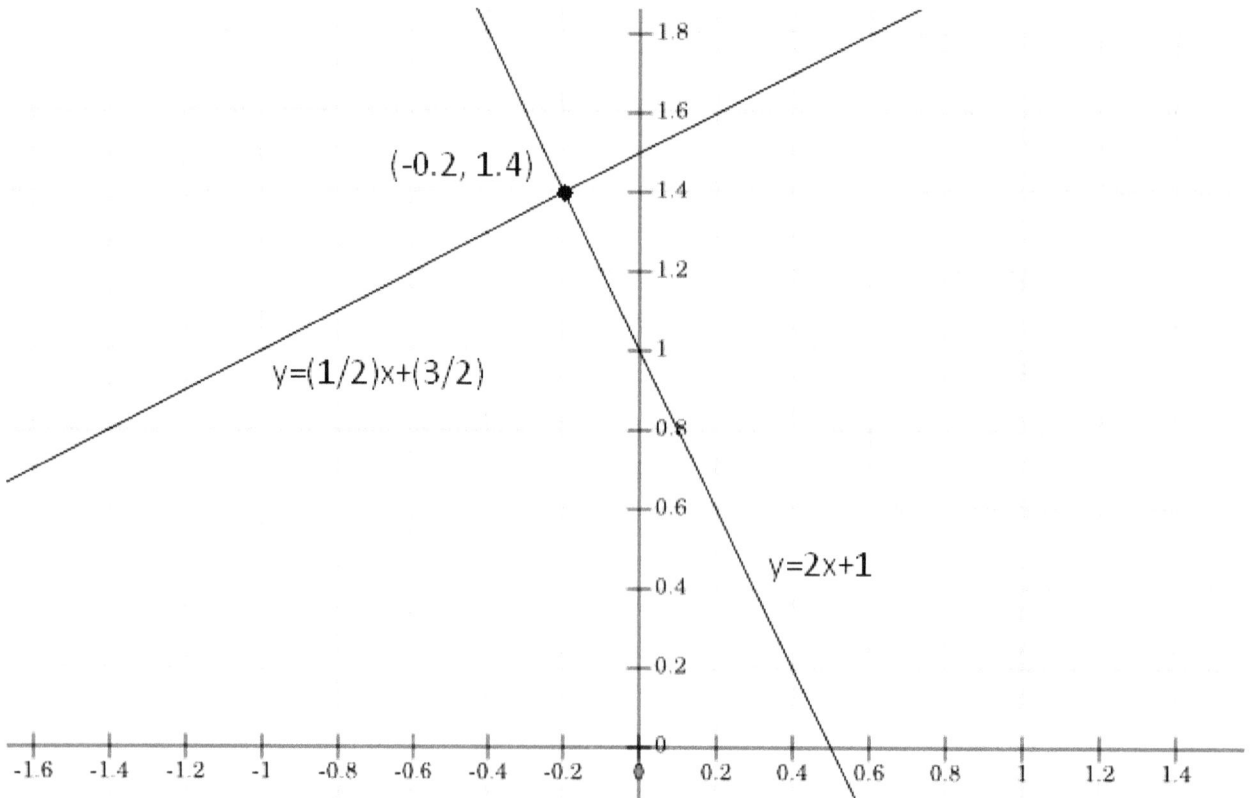

**Solution**

$2x + y = 1$

$-x + 2y = 3$

Calculate $x$ and $y$ intercept of $2x + y = 1$

When $x = 0$, $y = 1$

$Y$-intercept $(0, 1)$

When $y = 0$, $x = 1/2$

$X$-intercept $(0, \frac{1}{2})$

Calculate $x$ and $y$

intercept of $-x + 2y = 3$

When $x = 0$, $y = 3/2$

$Y$-intercept $(0, 3/2)$

When $y = 0$, $x = -3$

$X$-intercept $(0, \frac{1}{2})$

Point of intersection is

$(-0.2, 1.4)$

☞ <u>**Problem 1:**</u> **The length of a rectangle is twice than its width. If the perimeter of the rectangle is 52m. Find the dimensions of the rectangle?**

**Teacher Questions**

1. If the width is '$x$', what will be the length?

2. What is formula for the perimeter of the rectangle?

3. How will you incorporate the given length and width ratio to the perimeter formula?

4. Examine both the equations, what do you think is the better method to adopt, substitution method or graphical method?

### Solution

Let width of rectangle $= x$

length of rectangle $= y$

We know that length of rectangle is twice of its width, therefore,

$y = 2x$ ……. (i)

Also perimeter $= 2*($length + width$)$

$42 = 2*(x+y)$ …… (ii)

Substituting value of '$y$' from equation (i) in equation (ii),

$42 = 2*(x+2x)$

$21 = 3x$

$x = 7$

Put $x = 7$ in equation (i),

$y = 2x$

$y = 2(7) = 14m$

Therefore dimensions of rectangle are $7m$ and $14m$.

☞ **Problem 2:** The sum of two numbers is 40 and difference is 20. Write a system of equations and solve graphically. Also state whether the system is independent, dependent or inconsistent?

1. How will you represent the sum of two numbers in the form of equation?

2. Write an equation representing the difference of two numbers as 20?

3. How will you find $x$ and $y$-intercepts?

4. What is the difference between independent, dependent and inconsistent system?

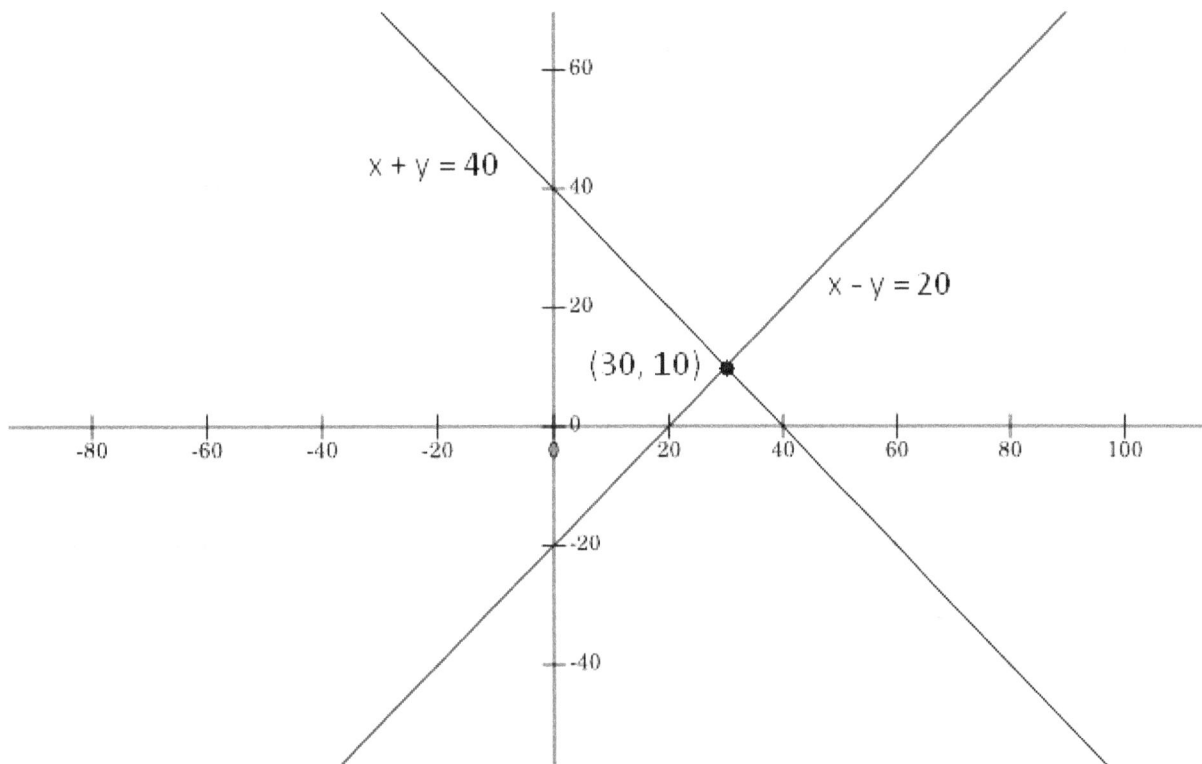

**Solution**

Let two numbers be '$x$' and '$y$'. According to the statement,

$x + y = 40$ ……. (i)

$x - y = 20$ …….. (ii)

Put $x = 0$ in eq. (i)

$y = 40$

$y$-intercept (0, 40)

Put $y = 0$ in eq. (i)

$x = 40$,

$x$-intercept (40, 0)

Join $x$ and $y$-intercept.

Put $x = 0$ in eq. (ii)

$y = 20$

$y$-intercept (0, 20)

Put $y = 0$ in eq. (ii)

$x = 20$

$x$-intercept (20, 0)

Join $x$ and $y$-intercept.

Intersection Point is (30,10)

Thus two numbers are 30 and 10.

System has one solution only, so it is an independent system.

## Video Suggestions

Please conduct a search on either YouTube or Teacher Tube to find appropriate videos for this lesson. Below are some suggested title searches:

➤ System of Linear Equations Part 1

➤ Solving Systems of Equations Using Substitution

➤ Systems of Equations (Substitution)

➤ Solving Systems of Equations Graphically

# Independent Instruction: Working On Your Own

☞ <u>Problem 1:</u> 12 oranges and 6 apples cost $30, whereas 8 oranges and 20 apples cost $52. What will be the cost per orange and cost per apple?

☞ <u>Problem 2:</u> For a fun show a total of 100 tickets were sold including regular tickets and child tickets. A child ticket costs $6 and a regular ticket costs $12. How many child tickets and regular tickets were sold if the revenue made was $960?

☞ <u>Problem 3:</u> Solve the system graphically and state whether the system is independent, dependent or inconsistent?

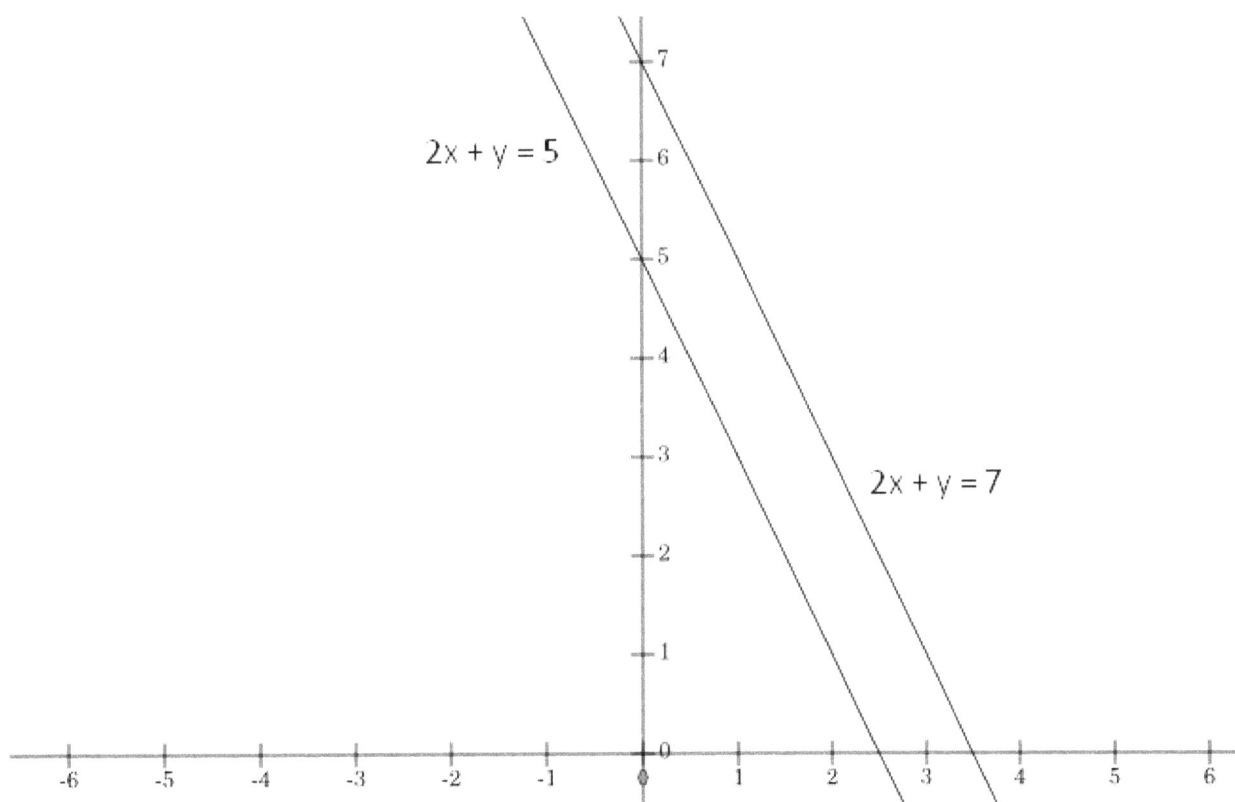

$$2x + y = 5$$

$$2x + y = 7$$

$2x + y = 5$

$2x + y = 7$

1. Let '$x$' be the cost per orange and '$y$' be the cost per apple.

$$12x + 6y = 30 \Rightarrow 2x + y = 5 \ldots\ldots\ldots \text{ (i)}$$
$$8x + 20y = 52 \Rightarrow 2x + 5y = 13 \ldots\ldots \text{ (ii)}$$

From equation (i),

$$y = 5 - 2x \ldots\ldots\ldots\ldots\ldots\ldots \text{ (iii)}$$

Substituting value of '$y$' in equation (ii),

$$2x + 5(5 - 2x) = 13$$
$$2x + 25 - 10x = 13$$
$$-8x = 13 - 25$$
$$-8x = -12$$
$$x = 12/8 = 1.5$$

Substitute $x = 1.5$ in equation (iii) to get the value of '$y$',

$$y = 5 - 2x = 5 - 2(1.5) = 5 - 3 = 2$$

Therefore, Cost per orange = $1.5

Cost per apple = $2

2. Let number of child tickets be '$x$' and number of regular tickets be '$y$'. Then,

$$x + y = 100 \ldots.. \text{ (i)}$$
$$6x + 12y = 960 \Rightarrow x + 2y = 160 \ldots.. \text{ (ii)}$$

From equation (i),

$$x = 100 - y \ldots.. \text{ (iii)}$$

Substituting the value of '$x$' in equation (ii),

$$x + 2y = 160$$
$$100 - y + 2y = 160$$
$$y = 160 - 100 = 60$$

Substitute $y = 60$ in equation (iii),

$$x = 100 - y = 100 - 60 = 40$$

Therefore, Number of regular tickets sold = 60

Number of child tickets sold = 40

3. The lines do not intersect each other, so the System is inconsistent.

# Mini-Assessment

1.  A system of linear equations is inconsistent if it has _____ solutions.

    **A.** Only 1  **B.** two  **C.** no solution

    **D.** infinite  **E.** None of Above

2.  Both the equations of a dependent system will have

    **A.** Same Slope  **B.** same slope and same $x$-intercept

    **C.** Same $x$-intercept  **D.** Same slope and same $y$-intercept

    **E.** Same $y$-intercept

☞**Problem 3:** The graph represents the following system of linear equations.

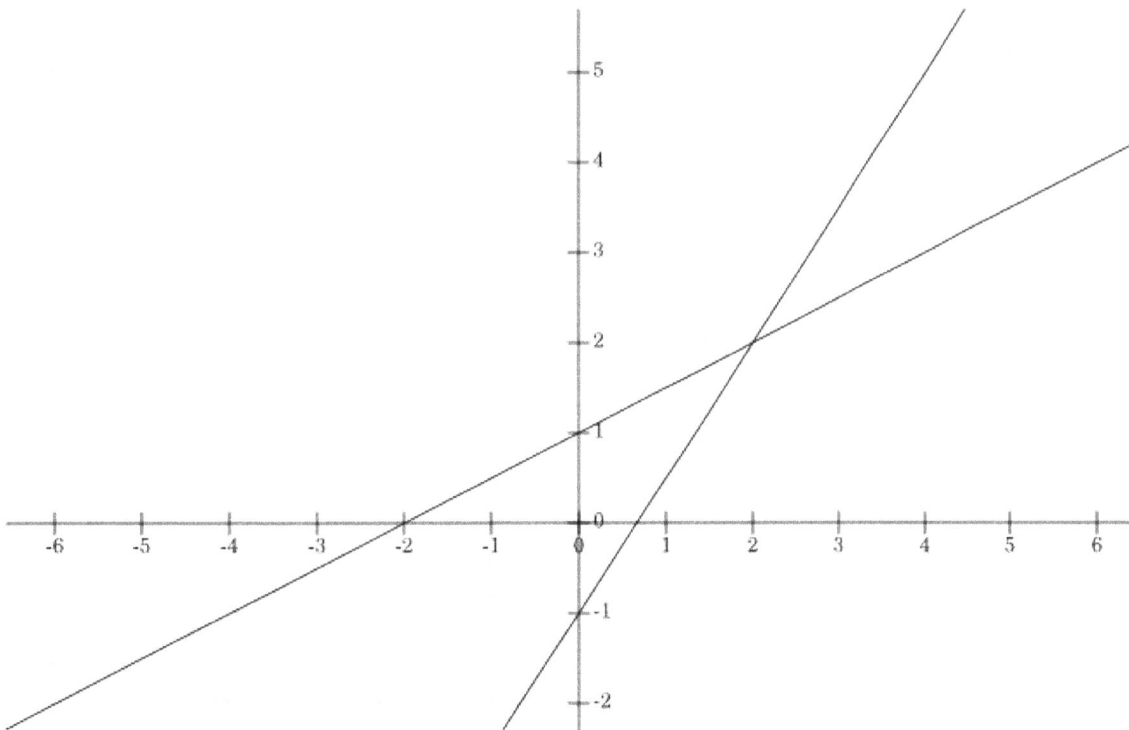

**A.** $3x + 2y = 2$
    $2y - x = 2$

**B.** $3x + 2y = -2$
    $2y + x = 2$

**C.** $3x - 2y = -2$
    $2y - x = 2$

**D.** $3x - 2y = 2$
    $2y - x = 2$

**E.** $-3x + 2y = 2$
    $2y - x = 2$

**4.** Solve the system of linear equations graphically.

$3x + y = 1$

$-x + 2y = 7$

**5.** State whether the given system is independent, dependent or inconsistent.

$(3/2)x + y = 1$

$3x + 2y = 7$

**6.** $A$ and $B$ are two supplementary angles. If $A$ is equal to 30 less than twice of $B$. Find measures of both the angles $A$ and $B$?

**7.** Solve the following system graphically and state whether it is independent, dependent or inconsistent?

$y = 2.5x + 3$

$2y = 5x + 6$

# Mini-Assessment Answers and Explanations

**1.** C

**2.** D

**3.** D

**4.** (Graph cannot be incorporated here in the notes section, therefore, just writing the answer)

The graph will show two intersecting lines meeting at point $(-1, 4)$.

Solution set is $(-1, 4)$

**5.** $(3/2) x + y = 1$ ---- (i)

$3x + 2y = 7$ ---- (ii)

Write eq. (i) in slope intercept form,

$Y = -(3/2) x + 1$ ---- (iii)

Write eq. (ii) in slope intercept form,

$2y = -3x + 7$

$y = -(3/2)x + 7$ --- (iv)

In eq. (iii) and (iv) slope is the same $(-3/2)$ where as y-intercept is different.

Therefore graphing these two equations will result into two parallel lines.

So the system will have no solution and it will be an inconsistent system.

**6.** $A + B = 180$ ---- (i)

$A = 2B - 30$ ---- (ii)

Substituting the value of $A$ from eq. (ii) in eq. (i),

$A + B = 180$

$2B - 30 + B = 180$

$3B = 180 + 30$

$3B = 210$

$B = 70$

Put $B = 70$ in eq. (ii),

$A = 2B - 30$

$A = 2(70) - 30$

$A = 140 - 30$

$A = 110$

Therefore both the angles measure 110 and 70.

**7.** The graph will represent two identical lines with same slope and same y-intercept.

System is dependent and it has infinitely many solutions.

## Lesson Reflection

### Solving Linear Equations Using Substitution Method

Substitution Method is useful, when in an equation, coefficient of a variable is 1.

### Steps to Solve Linear Equations Using Substitution Method

1.  Choose any one equation and calculate the value of one variable, let's say '$x$' in terms of '$y$'.
2.  Substitute the value of '$x$' in other equation.
3.  Solve this equation for the value of '$y$'.
4.  Replace back the value of '$y$' in the first equation to get the value of '$x$'.

### Solving Linear Equations Using Graphs

Graph both the equations one the plane. The point of intersection of both the lines will be the solution set.

### Types of Systems of linear Equations

If the system of linear equations has one intersection point, it has one solution and system is called independent system.

➤   If the system of linear equations has one line, it has infinitely many solutions existing on the whole line and this system is called a dependent system.

➤   If the system of linear equations has no intersection point, then it has no solution and system is known as inconsistent system.

# Lesson 7
# Linear Equation & Solutions Using Elimination, Matrices, and Determinants

## Lesson Description:

This lesson is designed to help students solve linear equations by using elimination, matrices, and determinants. Please be sure to utilize the questions to help spark student engagement and cover the vocabulary that is associated with this specific tutoring session. For your own knowledge, sample responses have been provided to guide you as well.

## Learning Objective(s):

In today's lesson, the learner will solve linear equations by using elimination, matrices, and determinants in 3 out of 4 trials with 75% or above accuracy.

## Introduction

As you may already know, linear equations and solutions are used to solve problems related to cost, interest, and principal amount. Another use of linear equations and solutions are to solve distance, rate, or time problems. For instance, individuals may try to find out how long it may take to travel to a certain city to another city. This is where system of linear equations would come into play to find out the time and distance of the trip.

## Questions to Engage Students

**1.** In your own words how would you describe the solution set of a system of linear equations?

**2.** A system of linear equations in two variables has two solutions. How many solutions will there be for a system of linear equations in three variables?

**3.** In your own words how will you define a square matrix?

**4.** Is it possible to calculate the determinant of matrices other than a square matrix?

## Connect Learning Objective(s) Student's Lives

System of linear equations is used:

**A)** in Mathematics and Calculus. For example, to find the equation of a conic, when two or three of its points are given.

**B)** To solve the problems related to Cost, Interest and Principal amount.

**C)** To solve Distance, Rate or Time Problems.

**D)** To solve geometric Applications, e.g finding dimensions of some geometric figure when perimeter and relation between the dimensions is given.

# Specific Vocabulary Covered

### Linear Equations

A linear equations is of the form $y = mx + b$ and it represents a straight line.

### Independent System

If the system of linear equations has one solution then it is called independent system.

### Inconsistent System

If the system of linear equations has no solution, then the system is known as inconsistent system.

### Dependent System

If the system of linear equations has infinitely many solutions, then this system is called a dependent system.

### Elimination Method

Elimination Method also called Addition Method is a method of solving system of equations by eliminating one variable, so that we are left with one variable equation.

## Matrix

A matrix is an array of numbers organized in rows and columns.

## Square Matrix

A square matrix has equal number of rows and columns.

## Augmented Matrix

A matrix obtained by combining the coefficients of the variables and the constants found in the system of linear equations is known as augmented matrix.

## Row Operations

Mathematical procedures applied on the rows of the matrix are called row operations. For example adding up the rows, multiplying the row by a constant etc.

# Direct & Guided Instruction: Modeling For You and Working With You

## Elimination Method

### Steps to Solve System Of Equations Using Elimination Method:

1. Eliminate one variable by adding or subtracting both equations.

2. Solve for the value of the variable that is left.

3. Substitute the value in any of the given equations, to get the value of second variable.

## Solving System of Equation Using Matrices

First we need to know about row operations of a matrix and reduced echelon form before going into the details of solving system of equations using matrices.

### Elementary Row Operations

1. Interchanging Rows.

2. Multiplying a row by a non zero number.

3. Adding a multiple of a row to another row.

### Reduced Echelon Form

1. All zero rows are at the bottom of the matrix.

2. The leading term in any non zero row is '1' and all terms above and below it are zero.

3. The leading '1' in any non-zero row occurs to the right of any leading '1' above it.

$$\begin{bmatrix} 1 & 0 & 0 & 2 \\ 0 & 1 & 0 & 1 \\ 0 & 0 & 1 & -1 \end{bmatrix}$$

## Solving System of Equation Using Matrices

### Steps to Follow

1. Write the system of equations in the form $Ax = B$, where $A$ is the matrix of coefficients, $x$ is matrix of variables and $B$ is the matrix of constants.

2. Write an Augmented matrix by combining matrix $A$ and $B$ together.

3. Apply row operations to the augmented matrix and convert it to reduced echelon form.

4. The last column shows the solution to the system.

## Solving System of Equation Using Determinants

### Cramer's Rule

1. Write the system as $AX = B$

$$\begin{bmatrix} a_{11} & a_{12} & a_{13} \\ a_{21} & a_{22} & a_{23} \\ a_{31} & a_{32} & a_{33} \end{bmatrix} \begin{bmatrix} x_1 \\ x_2 \\ x_3 \end{bmatrix} = \begin{bmatrix} b_1 \\ b_2 \\ b_3 \end{bmatrix}$$

2. Find $|A|$

**3.** Solution of the system can be found as,

$$x_1 = \frac{\begin{vmatrix} b_1 & a_{12} & a_{13} \\ b_2 & a_{22} & a_{23} \\ b_3 & a_{32} & a_{33} \end{vmatrix}}{|A|}, \quad x_2 = \frac{\begin{vmatrix} a_{11} & b_1 & a_{13} \\ a_{21} & b_2 & a_{23} \\ a_{31} & b_3 & a_{33} \end{vmatrix}}{|A|}, \quad x_3 = \frac{\begin{vmatrix} a_{11} & a_{12} & b_1 \\ a_{21} & a_{22} & b_2 \\ a_{31} & a_{32} & b_3 \end{vmatrix}}{|A|}$$

☞ **Problem 1:** **Solve the following system of equations using elimination method.**

$$-4x + y = 6$$
$$-5x - y = 21$$

### Teacher Questions

1. Why Elimination method is a better choice to use for this system?

2. How will you implement Elimination method?

### Solution

$-4x + y = 6$ ... (i)

$-5x - y = 21$ ... (ii)

Adding Eq. (i) and (ii) to eliminate '$y$',

$-9x = 27$

$x = -27/9 = -3$

Substitute '$x = -3$' in Eq. (i),

$-4x + y = 6$

$-4(-3) + y = 6$

$12 + y = 6$

$y = 6 - 12 = -6$

Hence, Solution Set is $(-3, -6)$

☞ **Problem 2:** **Solve the following system of equations using Matrices.**

$$x + 3y + 2z = 3$$
$$4x + 5y - 3z = -3$$
$$3x - 2y + 17z = 42$$

## Teacher Questions

1. How will you write the system as $AX = B$ using matrices?

2. How to write an augmented matrix?

3. To which matrix, row operations are applied?

## Solution

The augmented matrix can be written as,

$$\begin{bmatrix} 1 & 3 & 2 & \vdots & 3 \\ 4 & 5 & -3 & \vdots & -3 \\ 3 & -2 & 17 & \vdots & 42 \end{bmatrix}$$

Now, apply row operations to the augmented matrix to reduce it to reduced echelon form.

$$\begin{bmatrix} 1 & 3 & 2 & \vdots & 3 \\ 0 & -7 & -11 & \vdots & -15 \\ 3 & -2 & 17 & \vdots & 42 \end{bmatrix} \quad R_2 + (-4) R_1 \to R_2$$

$$\begin{bmatrix} 1 & 3 & 2 & \vdots & 3 \\ 0 & -7 & -11 & \vdots & -15 \\ 0 & -11 & 11 & \vdots & 33 \end{bmatrix} \quad R_3 + (-3) R_1 \to R_3$$

$$\begin{bmatrix} 1 & 3 & 2 & \vdots & 3 \\ 0 & -11 & 11 & \vdots & 33 \\ 0 & -7 & -11 & \vdots & -15 \end{bmatrix} \quad R_2 \leftrightarrow R_3$$

$$\begin{bmatrix} 1 & 3 & 2 & \vdots & 3 \\ 0 & 1 & -1 & \vdots & -3 \\ 0 & -7 & -11 & \vdots & -15 \end{bmatrix} \quad (-1/11) R_2 \to R_2$$

$$\begin{bmatrix} 1 & 3 & 2 & \vdots & 3 \\ 0 & 1 & -1 & \vdots & -3 \\ 0 & 0 & -18 & \vdots & -36 \end{bmatrix} \quad R_3 + 7 R_2 \to R_3$$

$$\begin{bmatrix} 1 & 3 & 2 & \vdots & 3 \\ 0 & 1 & -1 & \vdots & -3 \\ 0 & 0 & 1 & \vdots & 2 \end{bmatrix} \quad (-1/18) R_3 \to R_3$$

$$\begin{bmatrix} 1 & 0 & 5 & \vdots & 12 \\ 0 & 1 & -1 & \vdots & -3 \\ 0 & 0 & -1 & \vdots & 2 \end{bmatrix}$$

$R_1 + (-3) R_2 \rightarrow R_1$

$$\begin{bmatrix} 1 & 0 & 0 & \vdots & 2 \\ 0 & 1 & -1 & \vdots & -3 \\ 0 & 0 & -1 & \vdots & 2 \end{bmatrix}$$

$R_1 + (-5) R_3 \rightarrow R_1$

$$\begin{bmatrix} 1 & 0 & 0 & \vdots & 2 \\ 0 & 1 & 0 & \vdots & -1 \\ 0 & 0 & -1 & \vdots & 2 \end{bmatrix}$$

$R_2 + R_3 \rightarrow R_2$

Therefore, Solution set will be,

$x = 2, y = -1$ and $z = -2$

---

☞ **Problem 1: Solve the following system of equations using determinants.**

$$3x + y - z = -4$$
$$x + y - 2z = -4$$
$$x + 2y - z = 1$$

## Teacher Questions

1. What is the rule called used to solve the system of equations using determinants?

2. How will you get to know that there exists some solution to the system of equations?

3. What if $|A| = 0$?

**Solution**

The system of equations can be written as $AX = B$

$$\begin{bmatrix} 3 & 1 & -1 \\ 1 & 1 & -2 \\ -1 & 2 & -1 \end{bmatrix} \begin{bmatrix} x_1 \\ x_2 \\ x_3 \end{bmatrix} = \begin{bmatrix} -4 \\ -4 \\ 1 \end{bmatrix}$$

First of all, find $|A|$,

$$|A| = \begin{vmatrix} 3 & 1 & -1 \\ 1 & 1 & -2 \\ -1 & 2 & -1 \end{vmatrix}$$

$$|A| = (3)\begin{vmatrix} 1 & -2 \\ 2 & -1 \end{vmatrix} - (1)\begin{vmatrix} 1 & -2 \\ -1 & -1 \end{vmatrix} + (-1)\begin{vmatrix} 1 & 1 \\ -1 & 2 \end{vmatrix}$$

$$|A| = (3)[(1)(-1)-(2)(-2)]-1[(1)(-1)-(-1)(-2)]-(1)[(1)(2)-(-1)(1)]$$

$$|A| = (3)[-1+4]-1[-1-2]-(1)[2+1]$$

$$|A| = (3)(3)-1(-3)-(1)(3)$$

$$|A| = 9+3-3 = 9$$

$$x_1 = \frac{1}{|A|} \cdot \begin{vmatrix} b_1 & a_{12} & a_{13} \\ b_2 & a_{22} & a_{23} \\ b_3 & a_{32} & a_{33} \end{vmatrix}$$

$$x_1 = \frac{1}{9}\begin{vmatrix} -4 & 1 & -1 \\ -4 & 1 & -2 \\ 1 & 2 & -1 \end{vmatrix} = \frac{1}{9}\left[-4\begin{vmatrix} 1 & -2 \\ 2 & -1 \end{vmatrix} - 1\begin{vmatrix} -4 & -2 \\ 1 & -1 \end{vmatrix} - 1\begin{vmatrix} -4 & 1 \\ 1 & 2 \end{vmatrix}\right] = \frac{1}{9}[-4(-1+4)-1(4+2)-1(-8-1)] = \frac{1}{9}(-9) = -1$$

$$x_2 = \frac{1}{|A|} \cdot \begin{vmatrix} a_{11} & b_1 & a_{13} \\ a_{21} & b_2 & a_{23} \\ a_{31} & b_3 & a_{33} \end{vmatrix}$$

$$x_2 = \frac{1}{9}\begin{vmatrix} 3 & -4 & -1 \\ 1 & -4 & -2 \\ -1 & 1 & -1 \end{vmatrix} = \frac{1}{9}\left[3\begin{vmatrix} -4 & -2 \\ 1 & -1 \end{vmatrix} + 4\begin{vmatrix} 1 & -2 \\ -1 & -1 \end{vmatrix} - 1\begin{vmatrix} 1 & -4 \\ -1 & 1 \end{vmatrix}\right] = \frac{1}{9}[3(4+2)+4(-1-2)-1(1-4)] = \frac{1}{9}(9) = 1$$

$$x_3 = \frac{1}{|A|} \cdot \begin{vmatrix} a_{11} & a_{12} & b_1 \\ a_{21} & a_{22} & b_2 \\ a_{31} & a_{32} & b_3 \end{vmatrix}$$

$$x_3 = \frac{1}{9}\begin{vmatrix} 3 & 1 & -4 \\ 1 & 1 & -4 \\ -1 & 2 & 1 \end{vmatrix} = \frac{1}{9}\left[3\begin{vmatrix} 1 & -4 \\ 2 & 1 \end{vmatrix} - 1\begin{vmatrix} 1 & -4 \\ -1 & 1 \end{vmatrix} - 4\begin{vmatrix} 1 & 1 \\ -1 & 2 \end{vmatrix}\right] = \frac{1}{9}[3(1+8)-1(1-4)-4(2+1)] = \frac{1}{9}(18) = 2$$

☞ **Problem 2:** Using Elimination method, Solve the following system of equations.

$$4x + 4y + z = 24$$

$$2x - 4y + z = 0$$

$$5x - 4y - 5z = 12$$

## Teacher Questions

1. Why you think elimination method is a better choice to solve this system?

2. You need to get two equations in two variables using this system, how will you do that?

3. What will you do with these two new equations in two variables?

## Solution

$$x - 6y + 4z = -12 \ldots \text{(i)}$$

$$x + y - 4z = 12 \ldots \text{(ii)}$$

$$2x + 2y + 5z = -15 \ldots \text{(iii)}$$

Adding Eq. (i) and (ii),

$$2x - 5y = 0 \ldots \text{(iv)}$$

Multiplying Eq. (ii) by 2 and subtracting from Eq. (iii),

$$2x + 2y + 5z - 2(x + y - 4z) = -15 - 2(12)$$

$$2x + 2y + 5z - 2x - 2y + 8z = -15 - 24$$

$$13z = -39 \Rightarrow z = -3$$

Put $z = -3$ in Eq. (iii),

$$2x + 2y + 5(-3) = -15$$

$$2x + 2y - 15 = -15$$

$$2x + 2y = 0 \ldots \text{(v)}$$

Subtracting Eq. (iv) and (v),

$$2x - 5y - 2x - 2y = 0$$

$$-7y = 0 \Rightarrow y = 0$$

Put $y = 0$ in Eq. (v) to calculate value of $x$,

$$2x + 2(0) = 0 \Rightarrow x = 0$$

Solution Set is $(0, 0, -3)$

# Video Suggestions

Please conduct a search on either YouTube or Teacher Tube to find appropriate videos for this lesson. Below are some suggested title searches:

➤ Elimination Method

➤ Cramer's Rule to Solve a System of Equations

➤ Matrices and Determinants

➤ Solve Linear Equations with an Augmented Matrix

➤ Using Elimination Method

➤ Cramer's Rule

➤ Using Matrices

➤ Using Determinants

# Independent Instruction: Working On Your Own

## Questions

☞ **Problem 1:** Solve the system of equations using Matrices.

$$x - y + z = -1$$
$$4x + y - 2z = 5$$
$$-3x - 3y - 4z = -16$$

☞ **Problem 2:** Solve the following system of equations using determinants.

$$x - 3z = -17$$
$$-x - 2y - z = -14$$
$$-2x - y + 4z = -28$$

☞ **Problem 3:** Find solution to the system of linear equations using elimination method.

$$y - 2x = -3$$
$$x - y = 5$$

1. *Augmented Matrix is*

$$\begin{bmatrix} 1 & -1 & 1 & -1 \\ 4 & 1 & -2 & 5 \\ -3 & -3 & -4 & -16 \end{bmatrix}$$

$$\begin{bmatrix} 1 & -1 & 1 & -1 \\ 0 & 5 & -6 & 9 \\ 0 & -6 & -1 & -19 \end{bmatrix} \quad R_2 - 4R_1, R_3 + 3R_1$$

$$\begin{bmatrix} 1 & -1 & 1 & -1 \\ 0 & -1 & -7 & -10 \\ 0 & -6 & -1 & -19 \end{bmatrix} \quad R_2 + R_3$$

$$\begin{bmatrix} 1 & -1 & 1 & -1 \\ 0 & 1 & 7 & 10 \\ 0 & 6 & 1 & 19 \end{bmatrix} \quad -R_2, -R_3$$

$$\begin{bmatrix} 1 & 0 & 8 & 9 \\ 0 & 1 & 7 & 10 \\ 0 & 0 & -41 & -41 \end{bmatrix} \quad R_1 + R_2, R_3 - 6R_2$$

$$\begin{bmatrix} 1 & 0 & 8 & 9 \\ 0 & 1 & 7 & 10 \\ 0 & 0 & 1 & 1 \end{bmatrix} \quad \left(\frac{-1}{41}\right)R_3$$

$$\begin{bmatrix} 1 & 0 & 0 & 1 \\ 0 & 1 & 0 & 3 \\ 0 & 0 & 1 & 1 \end{bmatrix} \quad R_1 - 8R_3, R_2 - 7R_3$$

*Hence,* $x = 1, y = 3, z = 1$ *is the solution set.*

2. $A = \begin{bmatrix} 1 & 0 & -3 \\ -1 & -2 & -1 \\ -2 & -1 & 4 \end{bmatrix}, X = \begin{bmatrix} x \\ y \\ z \end{bmatrix}, B = \begin{bmatrix} -17 \\ -14 \\ -28 \end{bmatrix}$

*Lets find $|A|$ first,*

$|A| = 1 \begin{vmatrix} -2 & -1 \\ -1 & 4 \end{vmatrix} - 3 \begin{vmatrix} -1 & -2 \\ -2 & -1 \end{vmatrix} = 1(-8-1) - 3(1-4) = 1(-9) - 3(-3) = -9 + 9 = 0$

*As $|A| = 0$, therefore, there exist no solution to the system given.*

3. $\qquad y - 2x = -3 \dots$ (i)

$\qquad x - y = 5 \dots$ (ii)

Adding Eq. (i) and (ii),

$\qquad y - 2x + x - y = -3 + 5$

$\qquad\qquad -x = 2 \Rightarrow x = -2$

Put $x = -2$ in Eq. (ii),

$\qquad x - y = 5$

$\qquad -2 - y = 5$

$\qquad\quad -y = 5 + 2$

$\qquad\quad -y = 7 \Rightarrow y = -7$

Therefore, Solution set of the system is $(-2, -7)$.

# Mini-Assessment

☞ **Problem 1:** Which of the following is not a method of solving system of linear equations?

**A.** Addition Method      **B.** Matrices      **C.** Cramer's Rule

**D.** Graphing      **E.** Geometric Method

☞ **Problem 2:** Find $|A|$ for the following system of linear equations?

$$2x - 3y = 8$$
$$x + 2y = 5$$

**A.** 1      **B.** −1      **C.** 7      **D.** −7      **E.** 2

☞ **Problem 3:** Solve the following system of linear equations?

$$6x - 2y = 5$$
$$-3x + y = 7$$

**A.** (19/12, 9/4)      **B.** (0, 0)      **C.** System is consistent and independent.

**D.** System is consistent and dependent.      **E.** System is inconsistent.

☞ **Problem 4:** Solve the system of linear equations.

$$x + y = 6$$
$$8x + y = -8$$

☞ **Problem 5:** Solve the following system of linear equations Using Matrices.

$$3 + 2x - 7y = 0$$
$$-3 - 7y = 10x$$

☞ <u>**Problem 6**</u>: Solve the following system of equations Using Determinants.

$$x - 2y + z = -4$$
$$2x - 3y + 2z = -6$$
$$2x + 2y + z = 5$$

☞ <u>**Problem 7**</u>: Explain When it is best to use Elimination Method and Cramer's Rule?

# Mini-Assessment Answers and Explanations

**1.** E

**2.** C

**3.** E

**4.** $x + y = 6$ --- (i)

$8x + y = -8$ --- (ii)

Subtracting Eq. (i) and (ii),

$-7x = 14$

$x = -2$

Put $x = -2$ in Eq. (i),

$x + y = 6$

$(-2) + y = 6$

$y = 6 + 2 = 8$

**5.** $2x + 7y = -3$ ... (i)

$10x + 7y = -3$ ... (ii)

Using elimination method,

Subtracting Eq. (i) and (ii),

$8x = 0$

$x = 0$

Put $x = 0$ in Eq. (i)

$2x + 7y = -3$

$2(0) + 7y = -3$

$7y = -3$

$y = -3/7$

Solution Set $(0, -3/7)$

**6.** Using Cramer's Rule,

$|A| = [1(-3-4) + 2(2-4) + 1(4+6)] = -7 - 4 + 10 = -1$

$x = 1$

$y = 2$

$z = -1$

**7.** Elimination is best suitable to use when none of the coefficients are 1 or $-1$. Cramer's rule is best to use when determinant of matrix of coefficients is not zero.

# Lesson Reflection

## Elimination Method

1. Eliminate one variable by adding or subtracting both equations.

2. Solve for the value of the variable that is left.

3. Substitute the value in any of the given equations, to get the value of second variable.

## Elementary Row Operations

1. Interchanging Rows.

2. Multiplying a row by a non zero number.

3. Adding a multiple of a row to another row.

## Reduced Echelon Form

1. All zero rows are at the bottom of the matrix.

2. The leading term in any non zero row is '1' and all terms above and below it are zero.

3. The leading '1' in any non-zero row occurs to the right of any leading '1' above it.

## Solving System of Equation Using Matrices

1. Write the system of equations in the form $Ax = B$, where $A$ is the matrix of coefficients, $x$ is matrix of variables and $B$ is the matrix of constants.

2. Write an Augmented matrix by combining matrix $A$ and $B$ together.

3. Apply row operations to the augmented matrix and convert it to reduced echelon form.

4. The last column shows the solution to the system.

$Solving\ System\ of\ Equation\ by\ Cramer's\ Rule\ (Determinants):$

$AX = B$

$$\begin{bmatrix} a_{11} & a_{12} & a_{13} \\ a_{21} & a_{22} & a_{23} \\ a_{31} & a_{32} & a_{33} \end{bmatrix} \begin{bmatrix} x_1 \\ x_2 \\ x_3 \end{bmatrix} = \begin{bmatrix} b_1 \\ b_2 \\ b_3 \end{bmatrix}$$

$By\ Cramer's\ Rule,$

$$x_1 = \frac{\begin{vmatrix} b_1 & a_{12} & a_{13} \\ b_2 & a_{22} & a_{23} \\ b_3 & a_{32} & a_{33} \end{vmatrix}}{|A|}, \ x_2 = \frac{\begin{vmatrix} a_{11} & b_1 & a_{13} \\ a_{21} & b_2 & a_{23} \\ a_{31} & b_3 & a_{33} \end{vmatrix}}{|A|}, \ x_3 = \frac{\begin{vmatrix} a_{11} & a_{12} & b_1 \\ a_{21} & a_{22} & b_2 \\ a_{31} & a_{32} & b_3 \end{vmatrix}}{|A|}$$